CW00536237

UNKNOWN

SECRETS OF THE COSMOS FROM THE JAMES WEBB SPACE TELESCOPE

UNIVERSE

Published by Collins
An imprint of HarperCollins Publishers
Westerhill Road, Bishopbriggs, Glasgow G64 2QT
www.harpercollins.co.uk

HarperCollins Publishers
Macken House, 39/40 Mayor Street Upper,
Dublin 1, D01 C9W8, Ireland

© HarperCollins Publishers 2024
Text © Tom Kerss
Cover photograph © NASA, ESA, CSA, STScI. Back cover image extended and modified.

Collins® is a registered trademark of HarperCollins Publishers Ltd

A catalogue record for this book is available from the British Library

ISBN 978-0-00-871102-3

10 9 8 7 6 5 4 3 2 1

Printed in India

If you would like to comment on any aspect of this book, please contact us at the
above address or online.
e-mail: collins.reference@harpercollins.co.uk

UNKNOWN

RETS OF THE COSMOS FROM THE JAMES WEBB SPACE TELESCOP

UNIVERSE

TOM KERSS

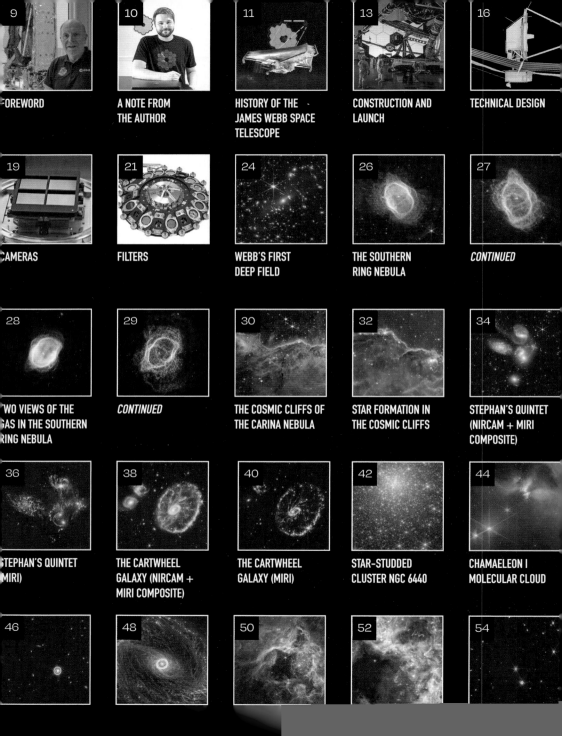

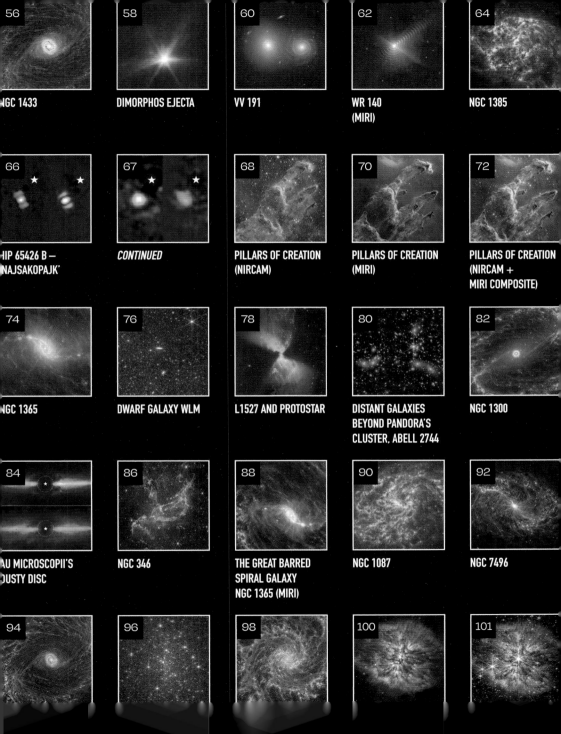

56
NGC 1433

58
DIMORPHOS EJECTA

60
VV 191

62
WR 140
(MIRI)

64
NGC 1385

66
HIP 65426 B —
'NAJSAKOPAJK'

67
CONTINUED

68
PILLARS OF CREATION
(NIRCAM)

70
PILLARS OF CREATION
(MIRI)

72
PILLARS OF CREATION
(NIRCAM +
MIRI COMPOSITE)

74
NGC 1365

76
DWARF GALAXY WLM

78
L1527 AND PROTOSTAR

80
DISTANT GALAXIES
BEYOND PANDORA'S
CLUSTER, ABELL 2744

82
NGC 1300

84
AU MICROSCOPII'S
DUSTY DISC

86
NGC 346

88
THE GREAT BARRED
SPIRAL GALAXY
NGC 1365 (MIRI)

90
NGC 1087

92
NGC 7496

94

96

98

100

101

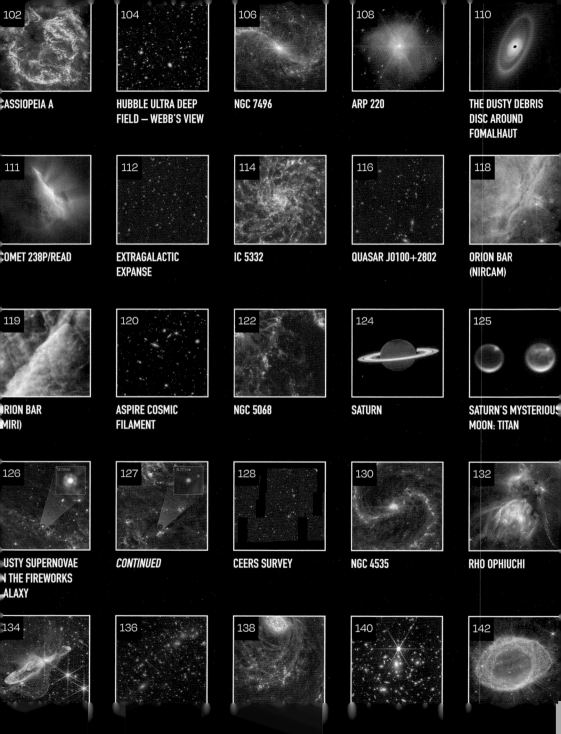

102 CASSIOPEIA A

104 HUBBLE ULTRA DEEP FIELD — WEBB'S VIEW

106 NGC 7496

108 ARP 220

110 THE DUSTY DEBRIS DISC AROUND FOMALHAUT

111 COMET 238P/READ

112 EXTRAGALACTIC EXPANSE

114 IC 5332

116 QUASAR J0100+2802

118 ORION BAR (NIRCAM)

119 ORION BAR (MIRI)

120 ASPIRE COSMIC FILAMENT

122 NGC 5068

124 SATURN

125 SATURN'S MYSTERIOUS MOON: TITAN

126 DUSTY SUPERNOVAE IN THE FIREWORKS GALAXY

127 CONTINUED

128 CEERS SURVEY

130 NGC 4535

132 RHO OPHIUCHI

134

136

138

140

142

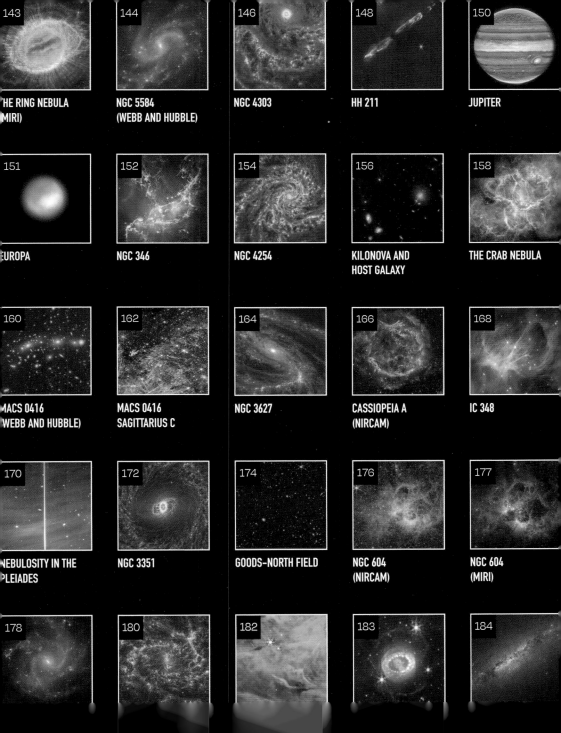

143
THE RING NEBULA
(MIRI)

144
NGC 5584
(WEBB AND HUBBLE)

146
NGC 4303

148
HH 211

150
JUPITER

151
EUROPA

152
NGC 346

154
NGC 4254

156
KILONOVA AND
HOST GALAXY

158
THE CRAB NEBULA

160
MACS 0416
(WEBB AND HUBBLE)

162
MACS 0416
SAGITTARIUS C

164
NGC 3627

166
CASSIOPEIA A
(NIRCAM)

168
IC 348

170
NEBULOSITY IN THE
PLEIADES

172
NGC 3351

174
GOODS-NORTH FIELD

176
NGC 604
(NIRCAM)

177
NGC 604
(MIRI)

178

180

182

183

184

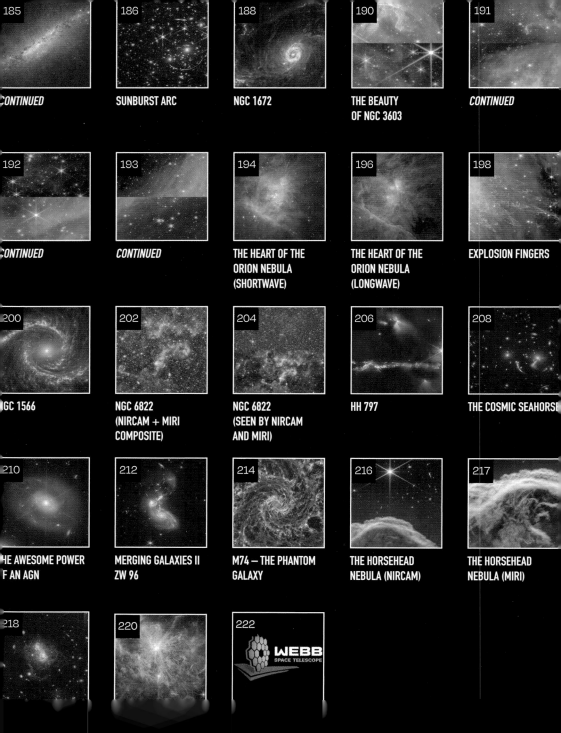

185 CONTINUED

186 SUNBURST ARC

188 NGC 1672

190 THE BEAUTY OF NGC 3603

191 CONTINUED

192 CONTINUED

193 CONTINUED

194 THE HEART OF THE ORION NEBULA (SHORTWAVE)

196 THE HEART OF THE ORION NEBULA (LONGWAVE)

198 EXPLOSION FINGERS

200 NGC 1566

202 NGC 6822 (NIRCAM + MIRI COMPOSITE)

204 NGC 6822 (SEEN BY NIRCAM AND MIRI)

206 HH 797

208 THE COSMIC SEAHORSE

210 THE AWESOME POWER OF AN AGN

212 MERGING GALAXIES II ZW 96

214 M74 — THE PHANTOM GALAXY

216 THE HORSEHEAD NEBULA (NIRCAM)

217 THE HORSEHEAD NEBULA (MIRI)

218

220

222 WEBB SPACE TELESCOPE

FOREWORD

Human eyes are the amazing outcome of hundreds of millions of years of evolution. They collect light through an adjustable iris, focus it onto the retina with a flexible lens, and then use special cells to convert the light into electrical signals. Our brains, which convert them into vibrant moving stereo images of the world around us, have also helped us understand that there are things the naked eye cannot see. Some objects are too faint, and others too distant to be seen in detail. We also know that there are invisible wavelengths of light that our eyes simply can't detect. Telescopes are the main tools we use to enhance our vision and help us understand our place in our Universe.

For more than 400 years, we have been improving telescopes to collect more light, see fainter, more distant objects, make sharper pictures, and reveal the invisible. We have discovered that we live on a small rocky planet orbiting an unremarkable star, in a galaxy containing hundreds of billions of stars, itself one of trillions of galaxies, many with supermassive black holes in their hearts, spread across a vast, expanding Universe that was born 13.8 billion years ago. The centuries-long development of this cosmic perspective is something that genuinely deserves the epithet 'awesome'.

Yet many questions remain. When did the first stars and galaxies form after the Big Bang and what did they look like? How have the galaxies evolved under the influence of mysterious dark matter and dark energy? How are new stars forming today from dense clouds of dust and gas? Do they host planets with atmospheres, hinting at the possibility of life beyond our Solar System?

Even before Hubble launched in 1990, astronomers began dreaming of a next generation space telescope capable of answering those questions: one with an enormous primary mirror to see extremely faint objects at infrared wavelengths, capable of detecting galaxies at high redshift and protostars through dust. It would need to be far from Earth, where it could get very cold and avoid the shimmer, glow, and absorption of our atmosphere. For decades, over 20,000 people at NASA, ESA, the CSA, in industry and in academia collaborated to create the astonishing machine called the James Webb Space Telescope.

On Christmas Day 2021, JWST rose above the jungle of French Guiana atop an *Ariane 5* rocket and soared towards its new home, 1.5 million kilometres from Earth. In July 2022, the first incredible scientific images and spectra were revealed from this new eye on the Universe. Since then, the discoveries have come rapidly, and in this book, Tom Kerss has gathered together some of the most stunning images and insights from JWST's first two years. This is only the beginning. JWST could operate for over twenty years, promising both new answers to old conundrums about the Universe, and surprising, unprecedented discoveries. The mission has taught us that when humans decide to work together, apply the best of our knowledge, expertise, and experience with dogged dedication over many years, we can achieve remarkable things. Given the mighty challenges facing us on our home planet, this is perhaps the most important lesson of all.

Dr Mark McCaughrean

Dr Mark McCaughrean is an Interdisciplinary Scientist on the JWST Science Working Group and an adjunct scientist at the Max-Planck-Institute for Astronomy in Heidelberg. He is the former Senior Advisor for Science & Exploration at the European Space Agency and is also the co-founder of the multi-award-winning Space Rocks. For more, visit markmccaughrean.net.

A NOTE FROM THE AUTHOR

We are living through a golden age of astronomy, in which giant super-telescopes and powerful space-based observatories have become icons. Their dazzling images of unimaginable scale and arresting beauty now permeate our culture. Who among us could deny the impact of the Hubble Space Telescope? Who hasn't had a space image or two as the wallpaper on their laptop or smartphone? We can all picture an observatory dome on the summit of a volcano above the clouds and the jewels of the Milky Way gleaming overhead. The final frontier has never been cooler, or closer.

In 2021, the James Webb Space Telescope left Earth, never to return, becoming a household name overnight. Its mission? To lift the veil on an unknown Universe. Webb has already peered deeply into space and time, uncovering the intimate secrets of shrouded protostars, the weather on worlds beyond the Solar System, and galactic events that shaped the evolution of the cosmos during its turbulent, formative epoch. In the course of its early studies, the telescope has provided images that have wowed astronomers and the general public alike. They have exceeded all expectations. Webb is an engine of wonder, and it will be remembered forever. Thus, it is with delight that I can present these images to you, the reader, as I believe they should be seen – in print. I hope they will challenge you, excite you and move you as much as they have me.

This book is dedicated to everyone involved in Webb's conception, creation, and operation. Your work has not only ushered in a new era of cosmic exploration, it has, more importantly in my opinion, brought inspiration and awe to billions of people. The images on these pages are among the most compelling, thought-provoking and humbling ever produced. They are marvellous works of art, made possible by your dedication and perseverance, and all of our lives are enriched by their existence. Thank you.

Tom Kerss is an astronomer and author based in the UK and US. Find out more about his books and other projects at tomkerss.co.uk.

HISTORY OF THE JAMES WEBB SPACE TELESCOPE

The origins of the James Webb Space Telescope (JWST) can be traced back to the late 1980s, but it was in 1994 when the *HST & Beyond* committee was formed at the Space Telescope Science Institute (STScI) to start seriously mapping out plans for a powerful successor to the iconic Hubble Space Telescope (HST). At the time, HST was still young and making ground-breaking observations, but it was designed and built in the 1970s with already ageing technology, and the astronomers on the committee recognised the need for a larger, more technologically advanced observatory, dedicated to detecting infrared radiation.

Detailed concepts for what had been called the Next Generation Space Telescope (NGST) emerged in 1996. Scientists determined that the primary mirror would need to be an unprecedented size to achieve fine resolution and extreme sensitivity at long wavelengths, well beyond the visible part of the electromagnetic spectrum. Initially, a segmented mirror with a 10-metre diameter was proposed. It was also thought that the observatory would benefit from being situated far from Earth for uninterrupted views of the sky, and would require a large sunshield to stay ultracool.

Early design work in the late 1990s focused on developing lightweight, yet durable, mirror technology. A silicon carbide layer with thin glass was proposed, but a novel design using gold-plated beryllium, which could maintain an extremely precise shape, to within one millionth of an inch, was ultimately selected. The mirror, which was now set to be 6.5 metres wide, would also be segmented into 18 hexagons to reduce cost and manufacturing time. Another key innovation was the advanced, five-layer sunshield that would deploy to block light and heat from the Earth, Sun, and Moon. This would allow JWST's mirrors and instruments to operate at cryogenic temperatures around -230 °C, no more than 50 degrees above absolute zero.

One of the biggest obstacles in the early planning stages was deciding where to locate NGST for optimal performance. Many ideas were evaluated, including orbits around the Earth at varying distances, all the way out to the Earth-Sun Lagrange Point 2 (L2) nearly one million miles away – four times further than the Moon. Scientists settled on L2 as the ideal space vantage point. NGST would be stationed in a halo orbit around this gravitational dead zone in space, in order to remain outside of the shadows of both the Earth and Moon. In this thermally stable environment, the dark side of the sunshield would experience minimal temperature changes, protecting the precise alignment of the mirror system.

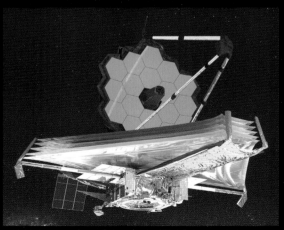

Artist's impression of the James Webb Space Telescope.

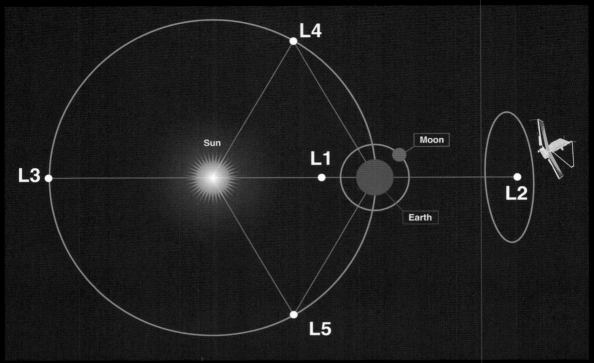

Webb orbits around a point in space called L2. Its halo orbit takes 168 days to complete.

In 2002, NGST was officially renamed the James Webb Space Telescope (JWST) after the former NASA administrator, James Edwin Webb (1906-1992), who led the US space programme through most of the 1960s, including the development of the historic Apollo missions. NASA's partners, the European Space Agency (ESA) and the Canadian Space Agency (CSA), signed on in 2004 and 2007 respectively, to contribute critical components and expertise as part of an international collaboration.

Throughout this period of the early 2000s, the mission's science goals, core technologies, and design requirements were continually refined as costs started to become a major concern. Perfecting the technology to precisely fold and stow the large, segmented mirrors and five-layer sunshield into a rocket payload for launch proved to be an immense engineering challenge. There were delays and major cost overruns, but the project persevered, surviving a proposed cancellation in 2011, as scientists worldwide eagerly awaited this next great leap in space-based observational astronomy.

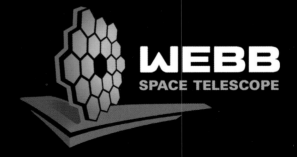

The official James Webb Space Telescope mission logo.

CONSTRUCTION AND LAUNCH

In 2003, TRW was awarded the contract by NASA to build the JWST. The company was acquired by Northrop Grumman the same year and became Northrop Grumman Space Technology. In 2004, fabrication of the telescope began in California. However, Webb's design could not be finalised until key technologies, critical to the project, had matured to a degree that were considered low risk to implement. Thus, the construction of components accelerated after this milestone, which was met in 2007. By 2011, the 18 mirror segments of Webb's iconic primary mirror were completed and tested, and in the years that followed, its science instruments – built in various locations – were delivered for evaluation and assembly. In 2012, the Mid-InfraRed Instrument (MIRI) was completed. In 2013, the Fine Guidance Sensor and Near Infrared Imager and Slitless Spectrograph (FGS-NIRISS) and Near-InfraRed Camera (NIRCam) arrived, as did the Near-InfraRed Spectrograph (NIRSpec.), The following year, all were installed into the Integrated Science Instrument Module (ISIM.)

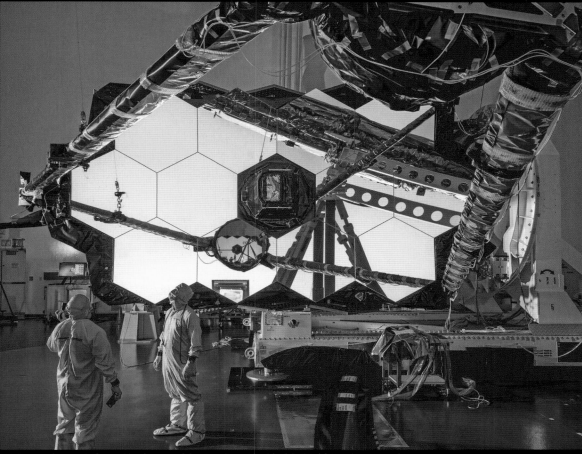

Engineers inspect the optical assembly structure that holds Webb's mirrors in place.
The reflection of the secondary mirror can be seen in the primary mirror.

Between 2015-2016, Webb's enormous primary mirror was assembled, along with its much smaller secondary and tertiary mirrors, and installed in the telescope. Cryogenic testing of the instruments and mirrors commenced in March 2016, and in November, engineers completed final assembly. Webb was ready. But not for launch. Not yet. First, the telescope had to undergo extensive testing. It experienced an anomaly in vibration testing which was swiftly resolved, but a much larger problem reared its head in 2018, when the sunshield was snagged and torn during a practice deployment. The launch was then delayed until 2020. Webb's launch vehicle and site had been selected back in 2005. It would be sent to space aboard an Arianespace *Ariane 5* on behalf of ESA, departing from the Centre Spatial Guyanais in Kuorou, French Guiana. When the selection was made, astronomers hoped that Webb would launch in 2013. By the time the telescope was assembled, the launch year had been revised to 2018 or 2019. As we approached the newly delayed launch window in May of 2020, the world faced a turmoil that no one could have predicted.

The global COVID-19 pandemic introduced huge challenges for scientists and engineers,

The finished and folded telescope prepares to leave Northrop Grumman in October 2021.

many of whom had to work on site to prepare Webb for launch. It was clear that the date would slip again. As the pandemic continued to cause chaos around the world, many felt it increasingly less likely that Webb could fly during a newly proposed 2021 window, and yet it did.

In a monumental effort, teams overcame further technical challenges, including with the launch vehicle. Webb – housed within the impressive *Ariane 5* – was rolled out to the launch pad on 23 December with a predicted launch date of 25 December, Christmas Day.

Webb's equivalent of a lens cap being removed.

The Ariane 5 *rocket carrying Webb blasted off from French Guiana on 25 December 2021.*

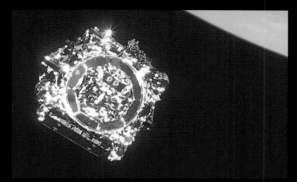

Webb, captured here from the Ariane 5 *upper stage, is released in orbit after launch.*

After decades of planning, development, construction, testing and heart-breaking delays, Webb was finally poised to ascend to the heavens. Now, the hopes and dreams of generations of astronomers depended on a successful launch. What they got was the greatest Christmas present imaginable.

At 12:20 UTC on 25 December 2021, Webb embarked on a flawless journey to space, beginning a voyage to its station nearly a million miles away. As it sailed out amongst the stars, astronomers flocked to their telescopes to capture it. The author acquired his own photos of Webb on its way on

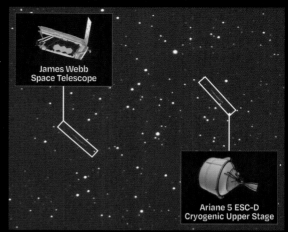

Integrated photography totalling 45 minutes of time, showing the trails of Webb and the Ariane 5 *upper stage sailing through the stars of Orion on 27 December 2021.*

27 December from Florida. Along the journey, the telescope underwent a series of nail-biting deployments, unfolding its solar arrays, sunshield, and mirror segments. A month after launch, on 24 January 2022, it entered its final orbit around L2, and spent months in commissioning. Its mirrors were aligned and tested, its instruments cooled and calibrated. On 11 July 2022, it was truly ready to start working.

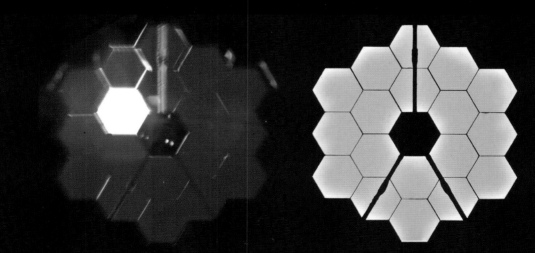

Using a special pupil, NIRCam was employed to capture 'selfies', allowing Webb's mirror segments to be aligned. The initial alignment (left) would have been unsuitable for providing sharp images. After many small adjustments, an excellent alignment (right) was achieved.

TECHNICAL DESIGN

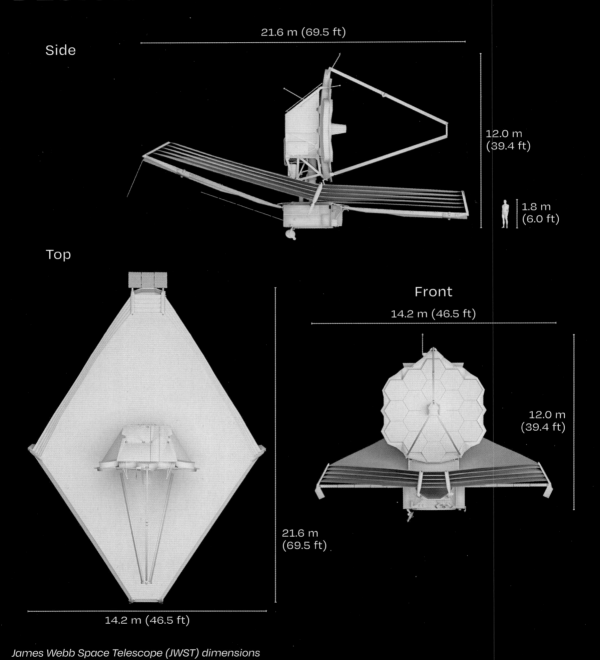

Side

21.6 m (69.5 ft)

12.0 m (39.4 ft)

1.8 m (6.0 ft)

Top

Front

14.2 m (46.5 ft)

12.0 m (39.4 ft)

21.6 m (69.5 ft)

14.2 m (46.5 ft)

James Webb Space Telescope (JWST) dimensions

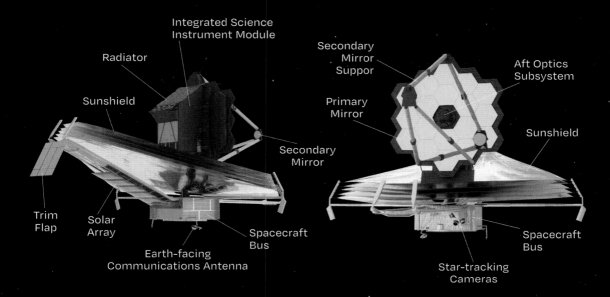

Integrated Science
Instrument Module

Radiator

Sunshield

Secondary
Mirror
Suppor

Primary
Mirror

Aft Optics
Subsystem

Sunshield

Secondary
Mirror

Trim
Flap

Solar
Array

Earth-facing
Communications Antenna

Spacecraft
Bus

Spacecraft
Bus

Star-tracking
Cameras

Webb represents a marvel of cutting-edge engineering and is one of the most, if not *the* most, ambitious space observatories ever constructed. Its innovative design incorporates numerous technical breakthroughs.

At the heart of the telescope lies its massive 6.5-metre-wide primary mirror. This critical component intercepts and focuses a wide band of infrared wavelengths. It is composed of 18 hexagonal segments, with surfaces made of gold-plated beryllium. They are highly thermally stable, minimising the need for Webb's optics to be readjusted over its mission lifetime.

The optical assembly also incorporates a similarly coated secondary mirror – much smaller by comparison, but still sizeable at 74 cm (29.1 inches) in diameter.
For comparison, the Spitzer Space Telescope, an earlier generation of infrared space observatory, has a primary mirror that is 85 cm (33.5 inches) wide. Infrared light is focused through an aperture in the centre of Webb's main mirror assembly and is reflected by a still-smaller tertiary mirror, on its way to the module containing the instruments.

A typical amateur telescope has a focal length of about one metre. Webb's folded design produces an effective focal length of 131.4 metres.

Sensitive observations in the infrared demand that the optical system and instruments operate at very cold temperatures. To achieve this, Webb's mirrors and instrument package are permanently shielded from the light of the Sun, Earth and Moon – the three brightest sources in its sky. As iconic as its segmented mirror is Webb's five-layer, diamond-shaped

An engineer inspects two segments of Webb's mirror. The one of the left has already been coated in gold, while the other has not.

DuPont Corporation – and is coated with aluminium. They are astonishingly thin – 0.05 mm (0.002 inches) for the first, Sun-facing layer and 0.025 mm (0.001 inches) for the other four layers. They disperse extremely efficiently such that the outermost layer may be 110 °C (230 °F) while the innermost layer is just -237 °C (-395 °F).

On the warm side of the sunshield is the spacecraft bus, containing communication and guidance systems, the solar array to generate solar power, and a curious protrusion called a trim flap, which is designed to balance the spacecraft as solar radiation exerts pressure on the sunshield. This obviates excessive wear on the spacecraft's reaction wheels, which would otherwise need to work harder to maintain precise pointing.

An engineer prepares the enormous, yet incredibly thin, Kapton sunshield for environmental testing in 2020.

sunshield. It measures 21.2 by 14.2 metres (69.5 by 46.5 feet), casting the telescope in shadow. Each layer is made of Kapton – a highly stable polyimide film developed by

Webb has a total mass of 6,200 kilograms, of which 705 kilograms is the main mirror. Its solar array generates 2,000 watts of power, and it beams data to Earth at a maximum rate of 28 megabits per second.

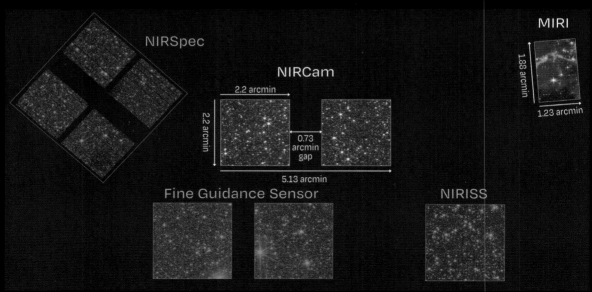

With its optical system aligned, Webb captured this focus test using all of its imaging instruments. NIRCam and MIRI are highlighted. The instruments do not overlap. Each focal plane covers a slightly different part of the sky in front of the telescope.

CAMERAS

T he ISIM houses four instruments:

- **Near InfraRed Camera (NIRCam)**
  Provided by the University of Arizona

- **Near InfraRed Spectrograph (NIRSpec)**
  Provided by ESA, with components from
  NASA and Goddard Space Flight Center

- **Mid-InfraRed Instrument (MIRI)**
  Provided by the European Consortium
  with ESA, and by NASA Jet Propulsion
  Laboratory

- **Fine Guidance Sensor / Near InfraRed**
  Imager and Slitless Spectrograph (FGS/
  NIRISS)
  Provided by CSA

Among them are the two scientific cameras responsible for Webb's most arresting imagery, NIRCam and MIRI, which have captured the data used to make the images in this book. Both cameras cover different regions of the infrared spectrum, and consequently offer different resolutions. As such, they support different kinds of scientific observations, though at times both may be used in concert for particularly detailed observations.

NIRCAM

NIRCam is Webb's primary imager, designed to capture extremely high-resolution images in the near-infrared wavelength range from 0.6 to 5 microns. Its array of detectors is divided into a shortwave channel (0.6 - 2.3 microns) and longwave channel (2.4 - 5.0 microns). One module houses eight detectors, and the other houses two for a total of ten. Each module is set on a dedicated focal

NIRCam (wrapped in protective film) after it arrived at NASA's Goddard Space Flight Center in 2013.

plane within the camera assembly, behind a complex optical system.

The detectors, built by Teledyne Technologies Incorporated, are made of mercury-cadmium-telluride (HgCdTe) and offer a resolution of

An array of four of the eight shortwave detectors that form one of NIRCam's imaging modules.

2040 x 2040 pixels. As such, the shortwave channel provides finer resolving power than the longwave channel. Micro-shutters enable precise targeting of astronomical sources by opening sets of shutters only over the objects of interest, while keeping others closed to minimise noise and detector exposure time.

NIRCam achieves a field of view of 2.2 x 2.2 arcminutes on each of its arrays. To achieve wide field vistas, images must be mosaiced.

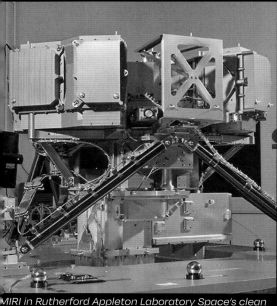

MIRI in Rutherford Appleton Laboratory Space's clean room in 2010.

MIRI

MIRI takes Webb's vision into the mid-infrared, with a sensitivity of 5 to 28 microns. This capability is key for piercing through clouds of dust and gas to reveal infant stars and their fledgling planetary systems. It also enables analysis of the oldest light in the Universe, which has been greatly redshifted by cosmic expansion. The camera contains three nearly

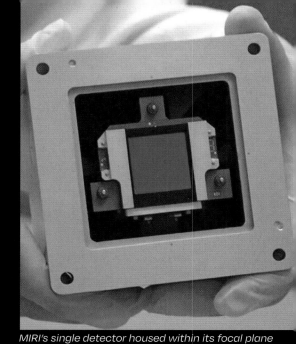

MIRI's single detector housed within its focal plane module.

identical detectors, made of arsenic doped silicon by Raytheon Vision Systems, of which one is used for imaging. The other two are used for spectroscopy. The detectors offer a resolution of 1024 x 1024 pixels, but only a portion is used to capture the images seen in this book.

For standard images, MIRI has a field of view of 1.2 x 1.9 arcminutes. For a given aperture, a telescope's resolving power is inversely proportional to the wavelength observed. At longer wavelengths, spatial resolution is lower. This, coupled with the smaller pixel count over a similar field of view, entails that MIRI images generally do not appear as sharp as NIRCam images. However, mosaic images from MIRI at arbitrary large resolutions can be truly spectacular.

FILTERS

NIRCam and MIRI are monochromatic detectors. They rely on precision filter wheels and filters to selectively observe specific infrared wavelengths. On NIRCam, there is one filter wheel and one pupil wheel for each channel. Each wheel holds 12 pupils or filters, for a total of 48 optical elements. Of these, 29 are colour filters, carefully fabricated using multi-layer coatings deposited on substrates, which provides selective transmission.

Filters are named according to their central wavelength and bandpass characteristics. For example, F070W denotes the filter that is centred on 0.704 microns, with a wide (W) bandpass. F323N denotes the filter that is centred on 3.237 microns, with a narrow (N) bandpass. NIRCam also has medium (M) filters and extra wide (W2) filters.

MIRI incorporates nine wideband filters for standard imaging. The shortest is centred on 5.6 microns; the longest on 25.5 microns.

The images in this book are typically created by combining multiple filters from NIRCam, MIRI, or both. Consider a regular camera, such as the camera on your smartphone. Its pixels measure the intensity of red, green and blue light. These values are combined to produce what appears to be a full-colour image. We can select three filters from Webb's cameras, and assign them the colours red, green and blue, even though none of them represent visible light. This will result in a false colour image. All infrared images are necessarily false colour, and the palette may be completely arbitrary. In many cases, more than three filters can be combined to create images with incredible depth and detail. In others, just one or two filters are used.

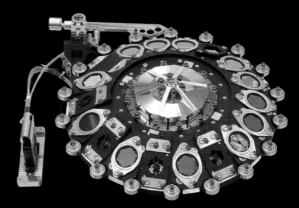

MIRI's filter wheel. Filters are rotated into the light path to control which wavelengths reach Webb's imaging detectors.

Occasionally, one filter is used to provide luminance (greyscale) data rather than colour, and some images combine Webb and Hubble data together. Generally, as it is with visible light, longer wavelengths are assigned warmer colours than shorter wavelengths, but the exact palette is a matter of choice for the person processing the image. In this regard, they are acting as an artist interpreting scientific data.

You will find a key accompanying each image on the following pages, showing which instruments and filters were employed and how colours were assigned to them. The book also includes images processed by the author, made from data captured by the telescope using the same methods as those published by NASA and ESA.

NIRCam	MIRI
F090W	F770W
F187N	F1130W
F212N	F1280W
F356W	
F405N	
F470N	

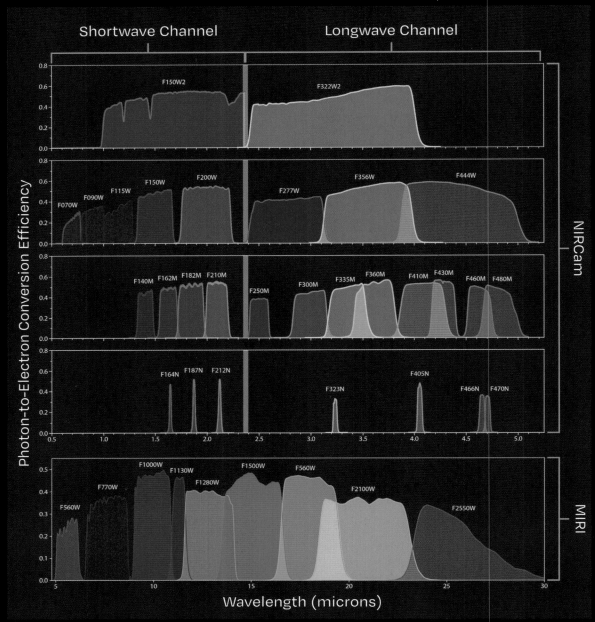

These graphs illustrate the bandpass characteristics of NIRCam and MIRI filters. NIRCam filters are divided between the shortwave and longwave channels. Photo-to-electron conversion efficiency is a measure of throughput for the entire telescope system, including the detectors. A theoretically perfect system would have a conversion efficiency of 1. The colours shown are suggested colours to apply to different filters when processing Webb images, however artists and scientists are free to use any colours palette they wish.

F277W calibrated

F277W calibrated and stretched

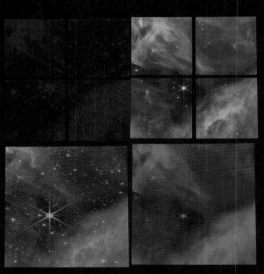

Stretched and aligned colour
channels:F150W (blue); F187N (cyan);
F277W (orange); F444W (red)

Colour composite of all four aligned
channels with cosmetic fixes

Monochrome image data is returned by the spacecraft. It is initially calibrated to remove flaws introduced by the instrument, such as noise and banding. Calibrated data is then stretched so a wider range of brightness values can be represented. Stretched images are aligned and assigned a colour. For balance and colour detail, a colour range that spans the visible spectrum is chosen. The final colour composite requires some manual cosmetic fixes, such as removing hot pixels, correcting saturated black spots in the centres of stars, and final contrast and colour balance adjustments.

WEBB'S FIRST
DEEP FIELD

Webb's first images stunned both astronomers and the wider public on 11 July 2022. Among them, the deepest and sharpest infrared image of the remote Universe that had ever been achieved. In just a tiny area of sky, no larger in apparent size than a grain of sand held out at arm's length, thousands of galaxies are revealed.

Central to the image is the galaxy cluster SMACS 0723. Fine details, such as star clusters and other features, are made clear in several of its galactic members. The cluster is so massive that it warps space and forms a giant 'gravitational lens', bending light itself to produce strange images of the background galaxies.

We see SMACS 0723 as it appeared 4.6 billion years ago, around the time our planet formed, whereas many of the background galaxies are considerably more distant still.

THE SOUTHERN RING NEBULA

Does this vista seem ghostly to you? In many ways, this is a real ghost – the spectre of a dying star that was once not dissimilar from the Sun. Our eyes are drawn to the bright star at the centre of the nebula, but it is a fainter companion – barely seen through the lower-left diffraction spike – whose death throes we are witnessing. The brighter star, its companion, exerts influence on the expanding cloud as the two orbit each other, sculpting delicate structures in its series of shells.

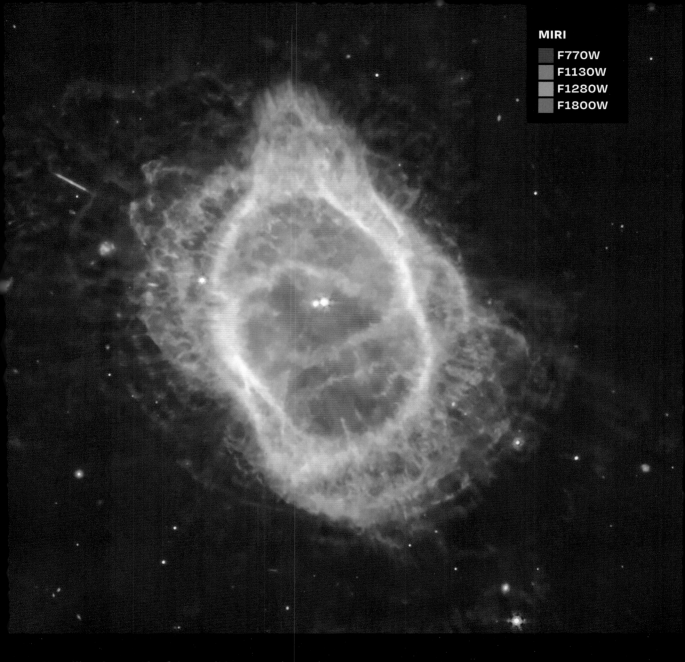

MIRI

	F770W
	F1130W
	F1280W
	F1800W

MIRI illuminates the fainter, dying star, shown here
in red. It is enshrouded by dust, which may one day
be ejected to become another layer of the planetary
nebula. At least eight layers have been released by
the star by the time we are witnessing it, thousands
of years into the process. The resulting nebula, about
2,000 light-years away, will continue to evolve and
transform for thousands more.

TWO VIEWS OF THE GAS IN THE SOUTHERN RING NEBULA

S elective combinations of NIRCam and MIRI channels allow astronomers to study different kinds of structures within the same object. The image on the left draws out hot gas near the centre of the nebula, which is pale in colour and relatively bright. It appears encompassed by a shell of cooler gas, shown in red, which can also be seen in the image on the right. This righthand image reveals cooler, molecular gas that was ejected at an earlier time. It has become bunched and braided as it expands, giving the impression of spokes.

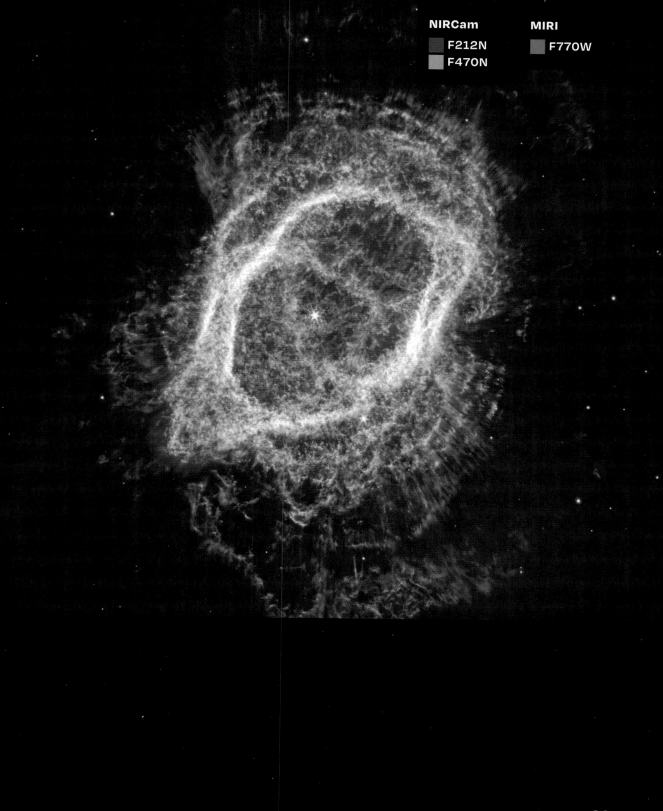

NIRCam
F212N
F470N

MIRI
F770W

THE COSMIC CLIFFS OF THE CARINA NEBULA

NIRCam

- ◼ F090W
- ◼ F187N
- ◼ F200W
- ◻ F335M
- ◼ F444W
- ◻ F470N

The Carina Nebula is home to innumerable curiosities. As a vast cloud of gas, constantly in flux at the mercy of powerful stars, it contains many regions of considerable interest. Here we see just one of them, called NGC 3324. It is a star-forming region, where young solar systems are beginning their lives, about 7,600 light-years from us.

What appears to be a coastline or mountain range, is in fact the edge of a vast cavity, which has been carved out by the stellar winds of powerful young stars (out of view from the upper area of the image). They shine fiercely in the ultraviolet, energising the gas and causing it to glow. So intense is their radiation, that the superheated gas gushes away from the cloud, giving the appearance of steam or sediment.

STAR FORMATION IN THE COSMIC CLIFFS

In this composite, the added data from MIRI allows us to peer more deeply into the Cosmic Cliffs, to unravel the secret lives of cocooned stars. Their formations are relatively fleeting events, taking at most about 100,000 years, and they are at their most elusive during this time. With MIRI, Webb has provided unprecedented opportunities to observe the process.

NIRCam

■ F090W
■ F200W
■ F444W

MIRI

■ F770W
■ F1130W
■ F1280W
■ F1800W

STEPHAN'S QUINTET
(NIRCAM + MIRI COMPOSITE)

A group of three galaxies drift together through the cosmos in this remarkably high-resolution composite image. Among them is a gatecrasher called NGC 7318B – one of a pair of interacting galaxies near the centre of the field. It has made quite an entrance, creating an enormous shockwave that is made visible in orange, courtesy of MIRI. In the chaos, countless new stars are being born in glittering starburst regions. The four galaxies form a cluster that lies about 290 million light-years away.

A fifth galaxy, NG7320 (shown on the left) is in the foreground – considerably closer to us at around 40 million light-years away. Its face-on spiral arms are rich with details, including individual stars, resolved by NIRCam's powerful detectors.

Studies of galaxy clusters, where closely bound members interact, are crucial to our understanding of the evolution of the Universe.

STEPHAN'S QUINTET
(MIRI)

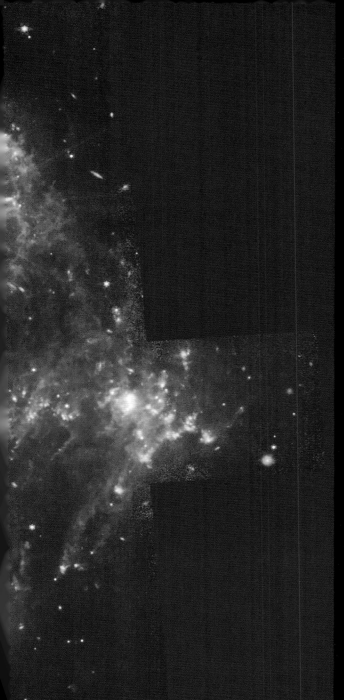

MIRI

F770W

F1000W

F1500W

MIRI's image of Stephan's Quintet is among the strangest produced by Webb in its first year, utilising three channels for a colourful result. Numerous red spots appear across the image, which are vastly more distant background galaxies. Within the group itself, the enormous gaseous structures that permeate each galaxy are exquisitely rendered. Gravity pulls them apart, creating tides and long tails. The pallid blue areas are relatively rich in hydrocarbons, which are essential to the chemistry of life, whereas the vibrant blue dots – particularly evident in NGC 7320 – are bright stars and star clusters not shrouded by dust. NGC 7319 appears at the top of the group, its active nucleolus blazing as a supermassive black hole, 24 million times more massive than the Sun, feeds in its core.

THE CARTWHEEL GALAXY
(NIRCAM + MIRI COMPOSITE)

NIRCam

- ■ F090W
- ■ F150W
- ■ F200W
- ■ F277W
- ■ F356W
- ■ F444W

MIRI

- ■ F770W
- ■ F1000W
- ■ F1280W
- ■ F1800W

Like a spinning firework 500 million light-years away, the Cartwheel Galaxy (PGC 2248) steals the show from its smaller companions in this composite image. Hydrocarbon-rich dust, which glows in the infrared, forms prominent rings around and spokes of pink and red, greatly enhanced by MIRI. About 400 million years ago, two galaxies collided at breakneck speed to produce the cartwheel's rings, which are shockwaves emanating from the centre. The spiral structure remains intact for now, but the Cartwheel Galaxy will continue to evolve in shape as the full ramifications of the collision emerge.

THE CARTWHEEL GALAXY
(MIRI)

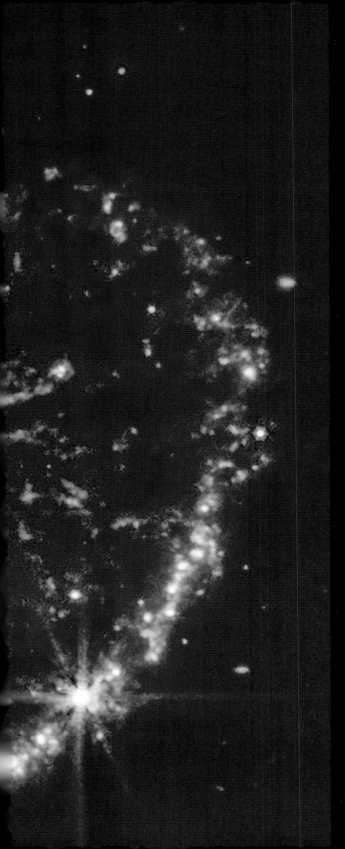

MIRI

▇	**F770W**
▇	**F1000W**
▇	**F1280W**
▇	**F1800W**

By assigning different colours to the four MIRI channels used to capture the Cartwheel Galaxy, we can measure the difference between its two prominent rings. The inner ring is hotter and will cool down as it expands. The outer ring is still active with the energy of young stars that have formed as a result of the shockwave. One of the nearby companion galaxies is also undergoing significant star formation, whereas the other is not.

The warm dust in these active galaxies is chemically rich, containing hydrocarbons and other compounds, including silicate dust. It is from this material that perhaps many new worlds will be born in the future.

STAR-STUDDED CLUSTER NGC 6440

L ike diamonds in the sky, tens of thousands of stars shimmer in this sharp NIRCam image of NGC 6440, a globular cluster roughly 28,000 light-years away. Such clusters are tightly packed, spherical swarms in which stars may be so close together that their separations are not measured in light-years, but light-months or light-days. If our planet were orbiting a star near the centre of NGC 6440, we would see potentially hundreds of stars in our daytime sky.

NIRCam
F115W
F200W
F277W
F444W

43

CHAMAELEON I
MOLECULAR CLOUD

A gaseous globule battles the flow of the surrounding nebula.

Roughly 630 light-years away, an infant protostar (left page, upper right) named Ced 110 IRS 4 illuminates a vast, wispy cloud of gas, bringing its delicate structures into view. This nebulous expanse is just one of several that collectively make up a large star-forming region, where many young stars are beginning their lives.

Hundreds of stars, usually obscured in visible light, are revealed as orange dots behind the cloud. Webb's instruments have studied their light to detect a remarkable diversity of ices present within the nebula. These simple compounds contribute to the compositions of exoplanets, offering clues about the possibility of life on other worlds.

THE URANIAN SYSTEM

The remote, ordinarily featureless, seventh planet comes alive in this sharp composite image from NIRCam, which reveals not only its elusive rings, but also 14 of its 27 known moons.

Galaxies galore populate the image field, teased out by Webb's extraordinarily powerful optics. Ordinarily, planetary images are composed of short exposures, resulting in pitch-black backdrops. In this case, long exposures remind us that no part of the sky is empty. Uranus appears suspended in front of the rich cosmic expanse of galaxies beyond it, like an ocean of island universes.

NIRCam

F140M
F210M
F300M
F460M

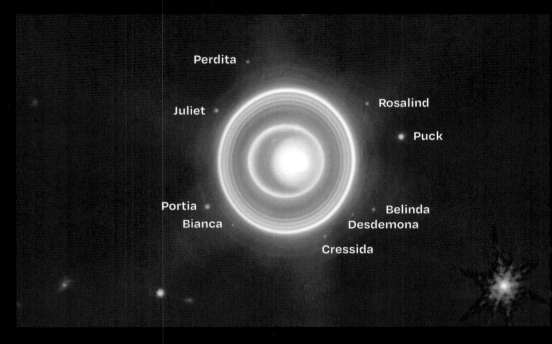

Uranus glows in the near-infrared, its polar gap a pearlescent white in this four-channel image. The extreme tilt of its polar axis is evident, as the Uranian north pole points broadly towards the inner Solar System. This provides us a near face-on view of its relatively dim ring system, which appears almost self-luminous to Webb's detectors.

It's fitting to see Uranus in the infrared, as both the planet and the radiation were discovered by the same scientist – English astronomer William Herschel. Uranus takes its name from the Greek god of the sky, but its many moons are named after characters from the works of William Shakespeare and Alexander Pope.

NGC 1512

Throughout this book you will find a collection of eye-catching spiral galaxies with a similar blue-orange palette. They were captured for the PHANGS (Physics at High Angular resolution in Nearby GalaxieS) survey and have been processed to draw out striking details in the spiral arms. Gas and dust are shown in orange and red, whereas stars are shown in blue. In the cores of these galaxies, we see large concentrations of older stars. Conversely, in the spiral arms we see younger stars. The youngest stars appear orange. They are still cocooned within dense nebulae. Star-forming regions are visibly distinct in the spiral arms, appearing as bright red spots. Beyond the galaxy at the focus of the image, look for distant background galaxies. Most appear as bright pink or blue sources.

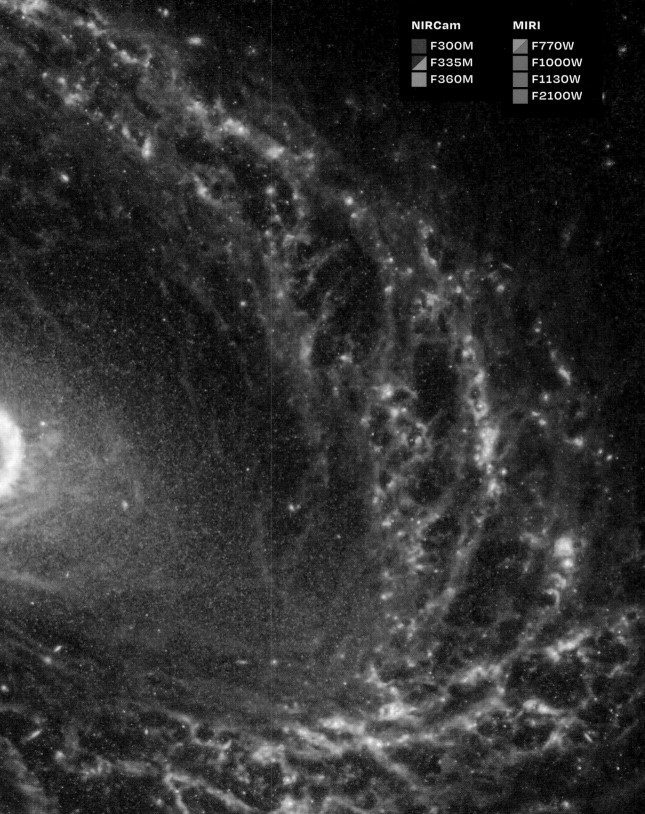

NIRCam

■ F300M
◥ F335M
■ F360M

MIRI

◥ F770W
■ F1000W
■ F1130W
■ F2100W

THE TARANTULA NEBULA

A grand vista unfolds across this spectacular mosaic, which is more than 300 light-years wide. We are peering into the turbulent heart of one of the largest known nebulae – the Tarantula – which is found not in the Milky Way itself, but in one of its satellite galaxies, known as the Large Magellanic Cloud. This puts the scene at about 170,000 light-years away from us.

NIRCam reveals tens of thousands of young stars for the first time, which were previously hidden out of view in dense clouds of dust. Massive young stars, including those in the prominent cluster NGC 2070, blaze in blue; while stellar infants, still buried in dust appear red. No visible light escapes these cosmic cocoons, but Webb's NIRCam is more than capable enough to scrutinise their inner workings.

The Tarantula is a majestic stellar nursery where countless more stars are yet to be born.

THE SPIDER'S HEART
(MIRI)

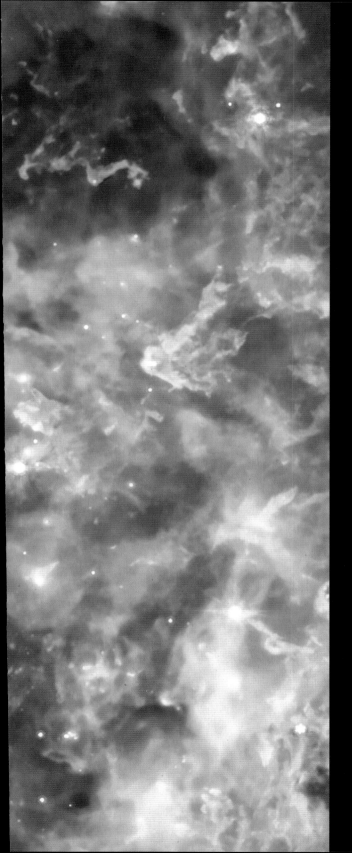

MIRI

░	F770W
▓	F1000W
▒	F1280W
▓	F1800W

MIRI's view of the Tarantula's core dulls the powerful young stars we see at shorter wavelengths, in favour of the glowing gases they excite with their intense radiation. We see textured layers of hydrocarbon-rich gas in blue and purple, and very dense clouds as dark regions, such as in the lower left of this image. Within the rugged cloudscape, cocooned stars that are totally obscured in visible light make themselves known as bright points.

THE NEPTUNIAN SYSTEM

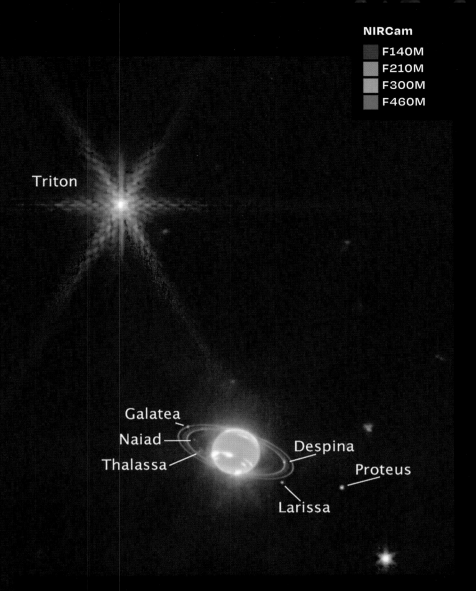

Triton

Galatea
Naiad
Thalassa
Despina
Proteus
Larissa

Neptune – that farthest planet from the Sun – seems to be drifting through intergalactic space in this NIRCam image. In fact, it is much closer to us than the many background galaxies strewn across the field. But as planets go Neptune remains astonishingly remote, and consequently, it's very cold. Despite this, the planet exhibits a surprising variety of weather, including high-altitude clouds of methane ice, which appear as bright spots in its southern hemisphere. Also revealed are the planet's faint rings, and seven of its moons.

The brightest of these by far is Triton to the upper left, which looks like a star casting, creating Webb's signature eight-way diffraction spikes. Triton is similar in size to Pluto, and may have originated from the Kuiper Belt before being captured into orbit by the Ice Giant world.

NGC 1433

NGC 1433 is a Seyfert Galaxy about 46 million light-years away. To find out more about this image, see page 48.

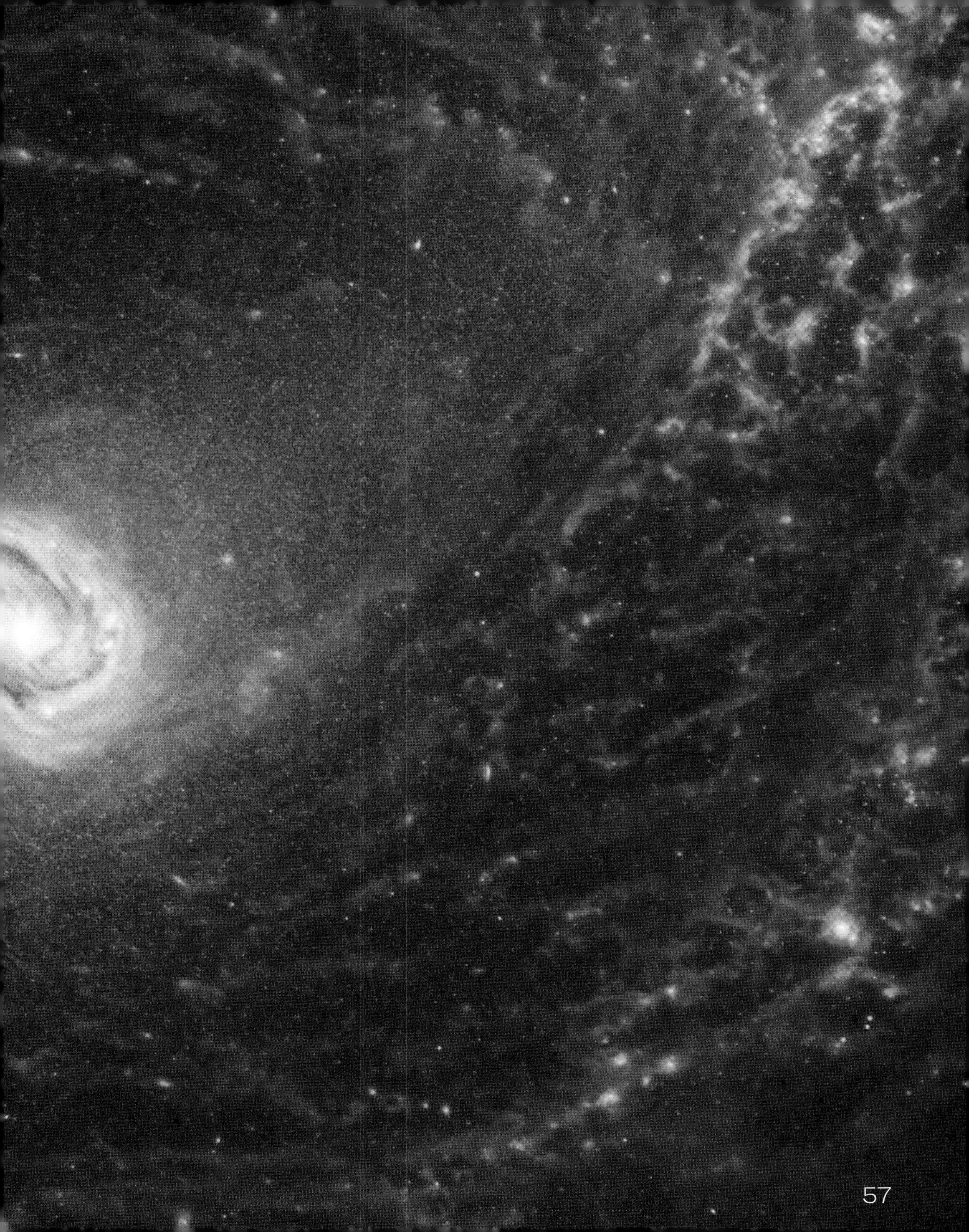

DIMORPHOS EJECTA

Not all of Webb's images are spectacular, varicoloured vistas. In some cases, the telescope's observations can seem unimpressive at first glance. However, there's always more than meets the eye. This image, captured with a single filter and assigned red only, appears to be a lone star immersed in spare dust, but it depicts something much more exotic. In September 2022, NASA carried out its remarkable Double Asteroid Redirection Test (DART) deliberately crashing a small spacecraft into Dimorphos, a minor planet moon orbiting the asteroid Didymos. The DART mission was designed to test the feasibility of altering an asteroid's orbit to protect our planet from a devastating impact.

DART struck Dimorphos at 6.6 kilometres per second, creating a plume of ejecta, which Webb was tasked to look out for. The image was part of a collection of observations made by Webb in concert with other telescopes, to help evaluate the success of the mission. Wisps of material can be seen streaming outwards from the tumbling rock.

VV 191

In what is surely one of the all-time greatest team-ups, the legendary Hubble Space Telescope and Webb both contributed data for this glorious portrait of a galaxy pair. To the left is an elliptical galaxy without spiral arms; to the right a face-on spiral galaxy. Both are prefect specimens of their respective classifications.

NIRCam's detectors reveal the extended, dusty, spiral arms in silhouette against the bright halo of the elliptical neighbour, confirming that the two are not interacting, despite their proximity to one another.

The background also contains numerous interesting galaxies well behind the dazzling duo, which are approximately 700 million light-years away.

WR 140
(MIRI)

MIRI

■ F770W
■ F1500W
■ F2100W

Wolf-Rayet stars are among the hottest and most luminous stars known. WR 140 is no exception, but even for its class it is extremely bright. That's because it forms a spectroscopic binary with a very close companion. Such binary systems cannot be resolved but are instead detected through the Doppler effect as they co-orbit one another. WR 140 and its powerful O-type sibling are locked in a dance that takes eight years to make one full cycle. It's thought that when they reach their closest approach, powerful tidal forces eject expanding shells of cosmic dust. The periodicity creates the impression of ripples on water.

At the distance to this system, 5,600 light-years, the image spans a width of about three light-years.

NGC 1385

NGC 1385 is a Spiral Galaxy about 30 million light-years away. To find out more about this image, see page 48.

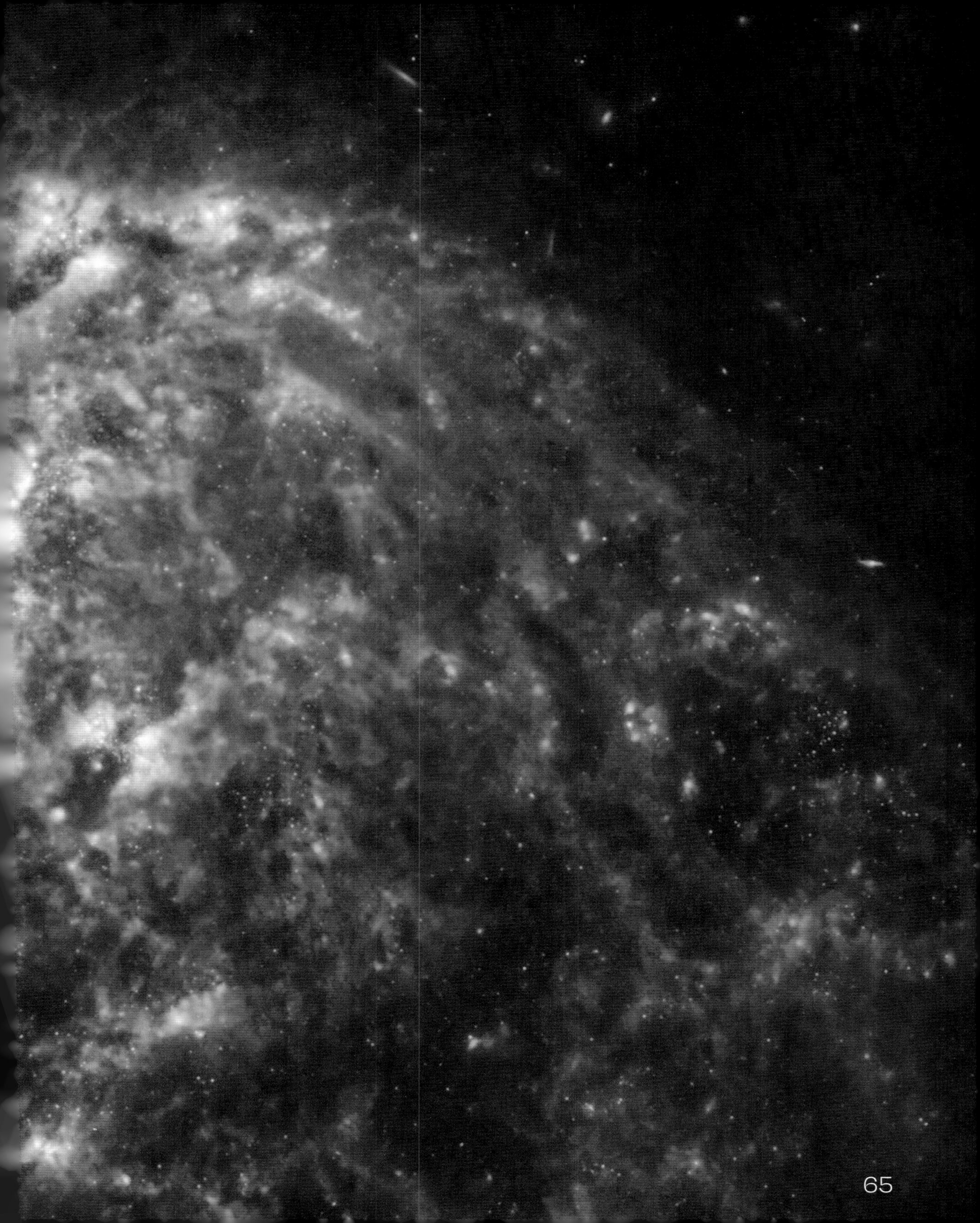

HIP 65426 B — 'NAJSAKOPAJK'

Four of NIRCam's filters were used to make these direct images of a world in another Solar System – an exoplanet. HIP 65426 is an unassuming star, 385 light-years away in the constellation of Centaurus. In 2017, astronomers discovered that it hosts a large gas giant exoplanet, dubbed a 'super-Jupiter'. In August 2022, it was granted the official name Najsakopajk, meaning 'Mother Earth' in the Zoque language, and in the following month it became the first exoplanet to be directly imaged by Webb.

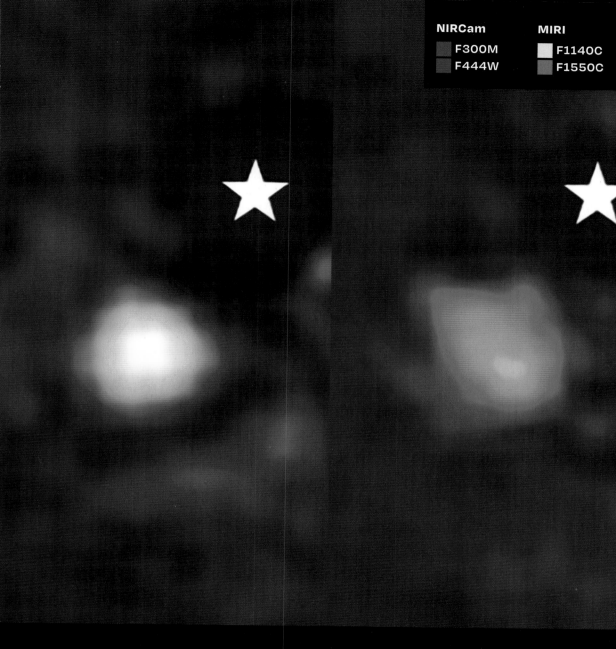

NIRCam

F300M
F444W

MIRI

F1140C
F1550C

Its host star, whose position is marked by the star-shaped icon, has been blocked out using a sophisticated mask called a coronagraph. This technology enables a clear image of the planet to be obtained.

While it's unlikely that such an exoplanet could harbour life, gas giants could have theoretically habitable moons. Webb's ability to directly image extrasolar systems makes it a powerful tool in the search for life in the Milky Way.

PILLARS OF CREATION
(NIRCAM)

The Pillars of Creation – the talons of the Eagle Nebula – have become an icon of astronomy, made famous by Webb's venerable predecessor, the Hubble Space Telescope. This sharp, infrared portrait raises the bar to new heights, showing us the Pillars as if with fresh eyes. The infrared filters make the towering columns of gas and dust seem semi-transparent, unveiling young stars within them. Truly infant stars appear as red dots.

Young stars often throw tantrums, emitting jets of energetic radiation and material that interact with the nebula. The colliding gas and dust may form a bow shock that radiates warmly, relative to its surroundings. In this image, shocks from supersonic jets have the appearance of molten rocks, glowing red from within.

69

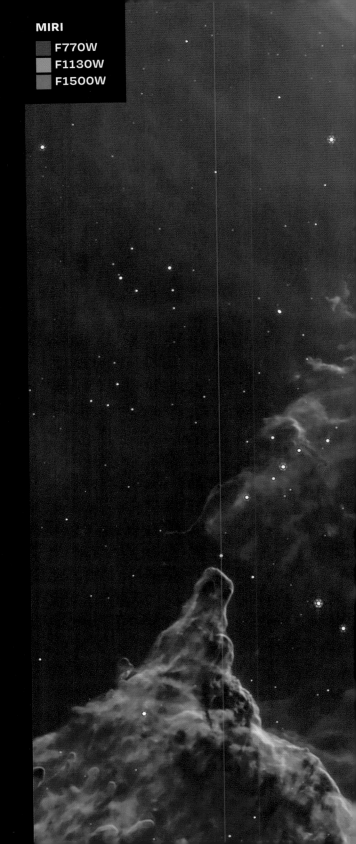

PILLARS OF CREATION
(MIRI)

MIRI's impression of the Pillars of Creation is wholly different and most spooky. The bustling metropolis seems almost deserted as only a few stars remain. MIRI's superpower is studying gas and dust. The young stars embedded within the Pillars don't shine brightly at such long wavelengths.

As its name suggests, this nebulous complex is a center of creation. Nebulae of this kind are destined to fragment and collapse, compelled by gravity. As they become more dense, relatively tiny fragments within the nebula heat up until they are hot enough for nuclear fusion to begin, ushering the births of new stars.

Webb is capable of profiling the density and composition of this dust with unprecedented confidence, helping astronomers to better understand the granular processes that precede and follow the formations of Solar Systems.

PILLARS OF CREATION
(NIRCAM + MIRI COMPOSITE)

This NIRCam and MIRI composite of the Pillars of Creation demonstrates the full power of the James Webb Space Telescope. Combining nine channels across the infrared spectrum, the image encompasses a staggering range of colour in a gorgeous palette.

NIRCam unmasks a menagerie of young stars, buried within the intricate pedestals of gas and dust that MIRI was purpose-built to investigate. Sitting about 6,500 light-years away, the nebulae and stars are physically local to one another, truly belonging in the same scene. Behind the Pillars, the tallest of which is nearly four light-years long, a great deal of gas and dust obscures the Universe beyond like a thick curtain.

While astronomical observatories are designed to generate scientific data, it is marvelous images such as this one that connect us to the cosmos. Among so many others, Webb's portrait of the Pillars of Creation will doubtless help to secure its place in our collective memory.

NGC 1365

NGC 1365 is a Barred Spiral Galaxy about 56 million light-years away. To find out more about this image, see page 48.

DWARF GALAXY
WLM

NIRCam

■ F090W
■ F150W
■ F250M
■ F430M

At first, it may appear that this is an ordinary image of stars within the Milky Way, complemented by a smattering of other galaxies beyond. In fact, almost all of the stars in this image are members of a faint dwarf galaxy about three million light-years from the Milky Way, known as Wolf-Lundmark-Melotte or WLM. We are looking straight through a portion of it, spanning approximately 1,700 light-years.

It is a testament to Webb's capabilities that so many stars are so finely resolved. Observations like this will give astronomers new insights into the lives of dwarf galaxies.

L1527 AND PROTOSTAR

NIRCam

F200W
F335M
F444W
F470N

A protostar rages at the heart of this dark cloud, feeding on the material at the core and growing in the process. As it does so, it ejects material in two distinct directions. It has carved two cone-shaped cavities into the cloud, the edges of which glow in the infrared. Subsequent ejections from the unborn star have further sculpted the cloud with strands and bead-like structures.

L1527 is about 460 light-years away in the constellation of Taurus.

DISTANT GALAXIES BEYOND PANDORA'S CLUSTER, ABELL 2744

F115W
F150W
F200W
F277W
F356W
F444W

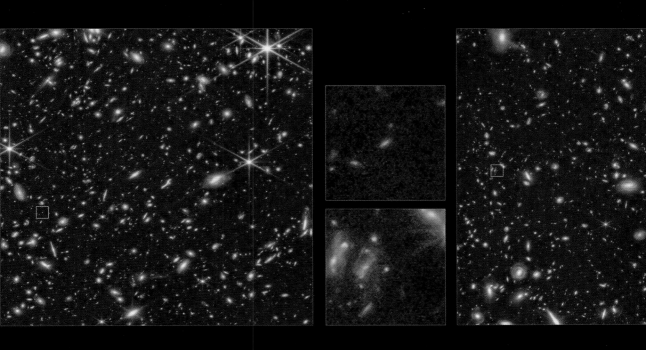

The scale of the scene presented in this seven-channel NIRCam image is almost impossible to comprehend. Some tens of thousands of galaxies, at a wide range of distances, populate the field. Pandora's Cluster is prominent, formed from the fuzzy, white galaxies that appear relatively bright. Beyond it, many more distant 'island universes' are drifting. Their images are warped by an immense 'gravitational lens', itself formed and figured by Pandora's Cluster. In some cases, the images of background galaxies are magnified, providing Webb with an opportunity to see in even more detail.

Among these faint sources are two of the most distant galaxies ever observed. One appears as it was just 450 million years after the big bang. The other, seen at an even earlier time, 350 million years after the big bang. They emerged as the first stars began to shine in the Universe, after a long period astronomers call the Dark Ages. Webb's extraordinary deep field images are astronomical treasure troves, overflowing with opportunities for discovery.

NGC 1300

NGC 1300 is a Barred Spiral Galaxy about 69 million light-years away. To find out more about this image, see page 48.

AU MICROSCOPII'S
DUSTY DISC

AU Microscopii (AU Mic) is a nearby red dwarf star, just 32 light-years away. In these NIRCam images, the star itself has been deliberately eclipsed by Webb's coronagraph, allowing deeper images to be taken, revealing a disc of dust around it. Astronomers have already discovered exoplanets within the disc, which appears edge on to us.

NIRCam's images present the disc in two wavelengths, with the blue channel appearing brighter than the red. The shorter wavelength represented in the blue image is preferentially scattered by finer dust, and the image suggests that this dust may be more abundant than expected. Webb is one of several world-class observatories engaged in studying this system, which may yet have surprises to offer.

NGC 346

NIRCam
	F200W
	F277W
	F335M
	F444W

Over 200,000 light-years away, embedded within the Small Magellanic Cloud, NGC 346 is a star cluster surrounded by a wispy nebula. NIRCam provides a detailed view of the hydrogen within the cloud, separating energetic hot (10,000 °C) hydrogen in pink and cold (-200 °C) molecular hydrogen in orange. The cloud is also rich in dust, which has implications for star systems that form within it. This cosmic dust contains the material necessary for the creation of planets.

Intricate features within the nebula indicate the presence of newborn stars, whose stellar winds are eroding the surrounding gas, carving ridges that show sharp contrasts.

THE GREAT BARRED SPIRAL GALAXY NGC 1365

(MIRI)

Massive star clusters and pockets, and glowing gas populate the face of NGC 1365, as made evident by MIRI's sensitive detector. Despite being 54 million light-years away, Webb is capable of teasing out fine details near the brilliant core of the galaxy for the first time. These include young star clusters on the edges of the 'bar' that extends to either side of the central core.

Bright knot-like features within the spiral arms tell of ongoing star formation throughout the disc, concentrated in a thick ring, while a population of older clusters orbit farther out. Webb's findings offer clues about how the orbits of stars depend upon where they form within the galactic disc.

F1130W
F2100W

NGC 1087

NGC 1087 is a Spiral Galaxy about 80 million light-years away. To find out more about this image, see page 48.

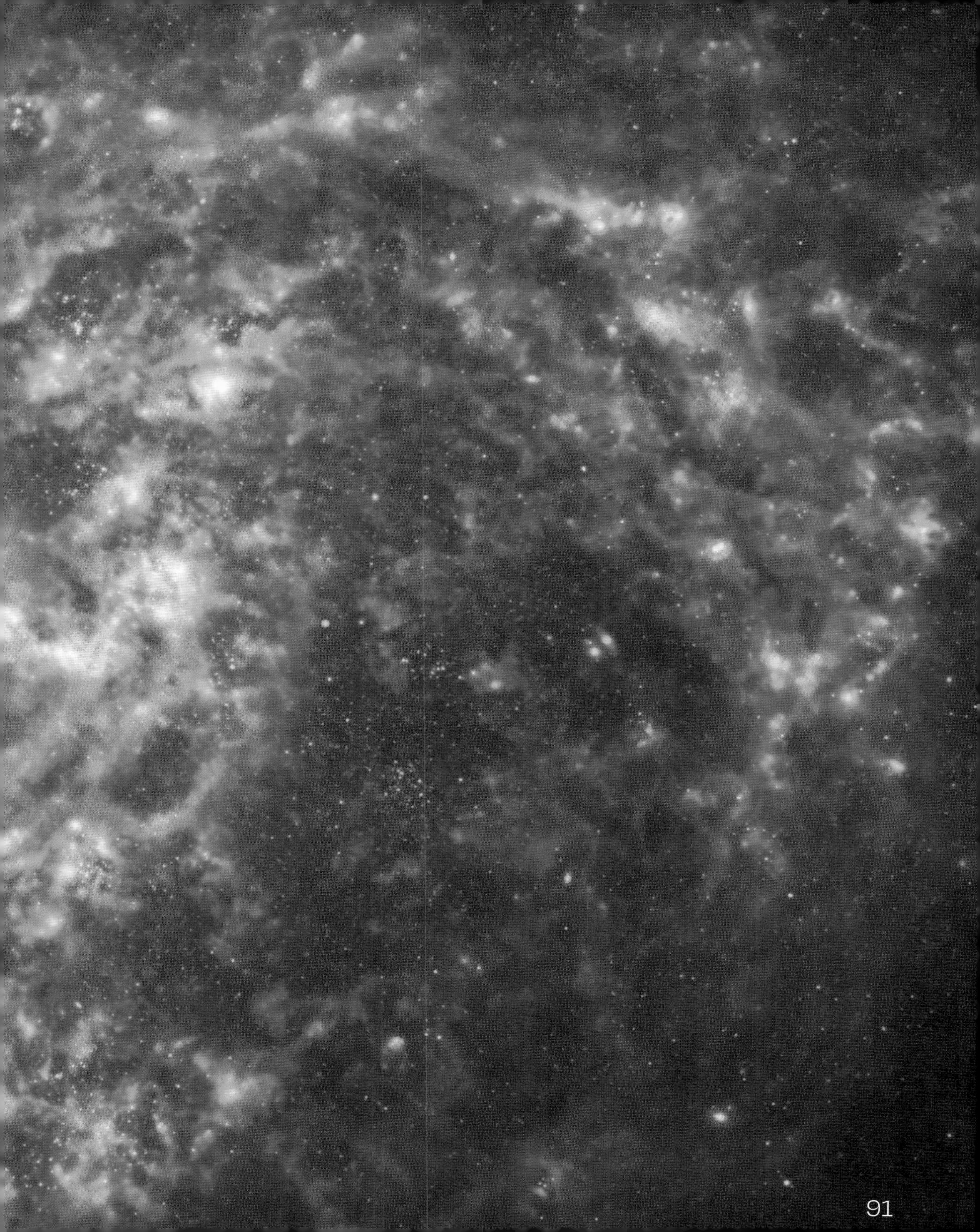

NGC 7496

92

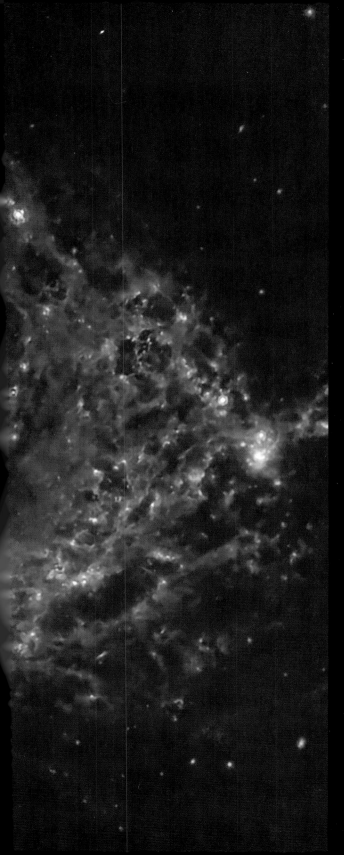

MIRI

■ F770W
■ F1000W
■ F1130W
■ F2100W

This cosmic portrait was captured by MIRI, as part of a programme to study the effects of stars and star clusters in nearby galaxies. Glowing cavities in nebulous regions pockmark the spiral arms where powerful young stars are energising and shaping the gas. The data revealed 60 candidate star clusters not previously known to astronomers, which are thought to contain the most recently formed star systems in the galaxy.

The lanes of chemically rich dust that trace out the spiral arms contain polycyclic aromatic hydrocarbons, whose signatures are detectable in the infrared. These complex organic molecules are crucial to the formation of new planets.

The core of NGC 7496 is home to a supermassive black hole, which is feeding on the surrounding material, creating a bright, active galactic nucleus (AGN). It emits high energy jets, and shines brilliantly enough to form diffraction spikes on MIRI's detector, despite being 24 million light-years away.

NGC 1433

While every part of this barred spiral galaxy is fascinating to explore, the eye is drawn to its core, where an unusually bright double ring feature illuminates MIRI's pixels. The galaxy's spiral arms have become tightly wound here, causing the central bar to appear bloated as an oval shape. Webb is nevertheless able to discern numerous bubbles within the central rings, where powerful stars are blowing away the interstellar medium and energising it in the process. This is a remarkable observation for a galaxy over 46 million light-years away.

As a Seyfert galaxy, NGC 1433 harbours a ravenous supermassive black hole, which is accreting and consuming material at a very high rate.

GLOBULAR CLUSTER M92

Two views from NIRCam show us the dazzling edges of the globular cluster M92. The clusters core, significantly brighter, was deliberately omitted, being far too bright to capture in detail with the desired exposure time. Instead, these fields provide insight into the lives of the stars within the cluster's suburbs. M92 drifts outside the Milky Way's disc, over 26,000 light-years away in the constellation of Hercules.

The compositions, ages and dynamics of its stars are of particular interest to astronomers, and Webb will provide new insights with its precision optics. M92 is roughly 100 light-years wide and home to several hundred thousand stars.

NGC 628

NGC 628, also known as M74, is a Spiral Galaxy about 32 million light-years away. To find out more about this image, see page 48. To see this galaxy in an alternative palette, see page 214.

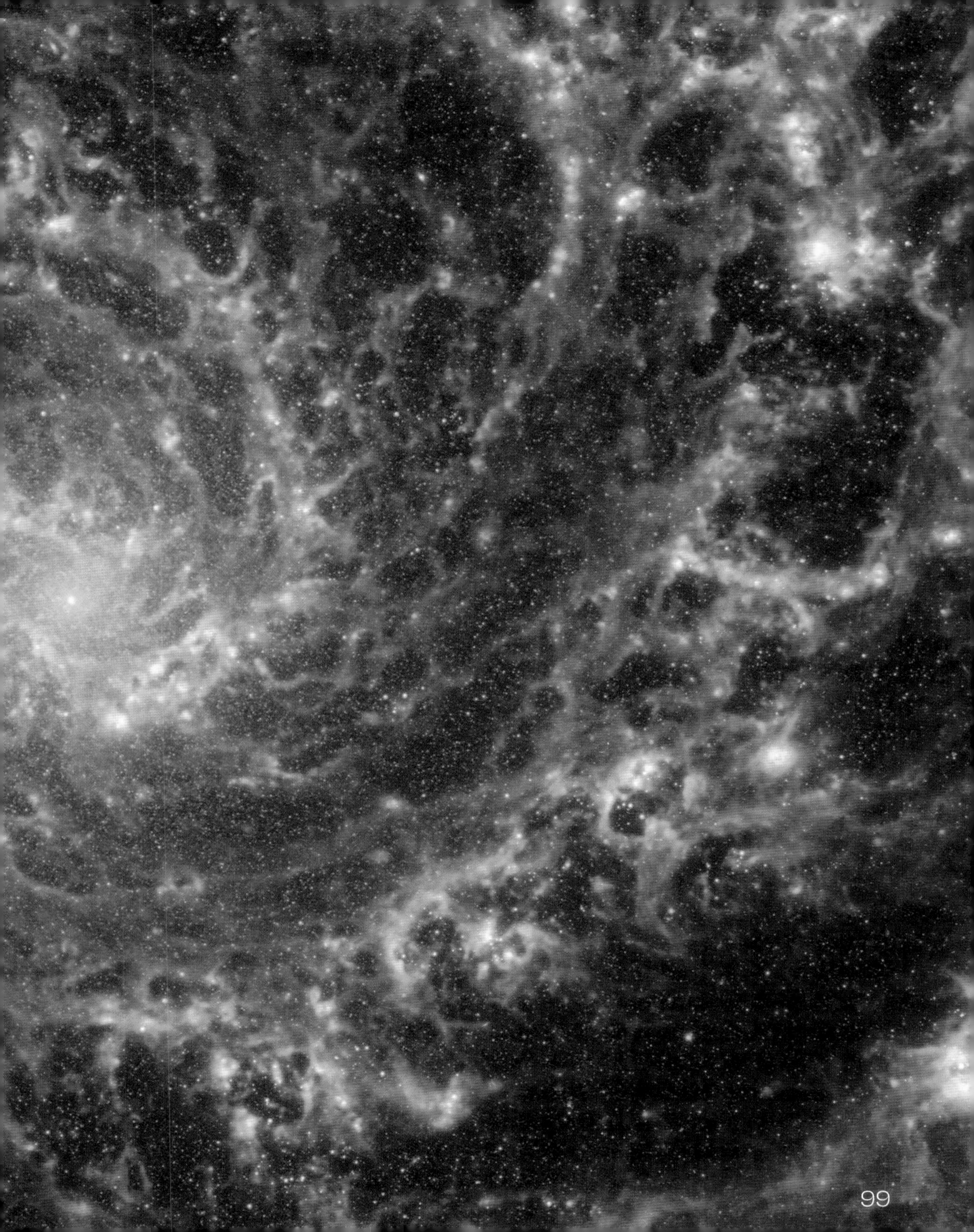

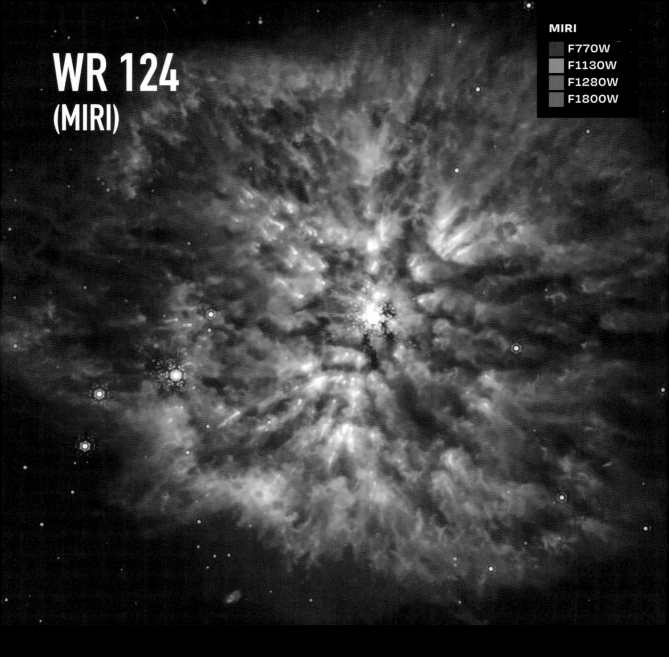

WR 124
(MIRI)

These side-by-side images of WR 124 illustrate the different capabilities of MIRI and NIRCam. As a Wolf-Rayet star, WR 124 produces and ejects large quantities of dust, which MIRI is perfectly suited to detect. A 10-light-year-wide nebula is shown expanding outward as the central star undergoes sporadic outbursts. Meanwhile, NIRCam completes the scene with a field of stars and background galaxies to frame the nebula.

(NIRCAM + MIRI)

NIRCam
- F090W
- F150W
- F210M
- F335M
- F444W
- F470N

MIRI
- F770W
- F1130W
- F1280W
- F1800W

WR 124 is in its final stages, approaching a cataclysmic death in the form of a supernova. Thanks to Webb, astronomers can unravel previously unsolvable mysteries about the final chapters of such influential stars.

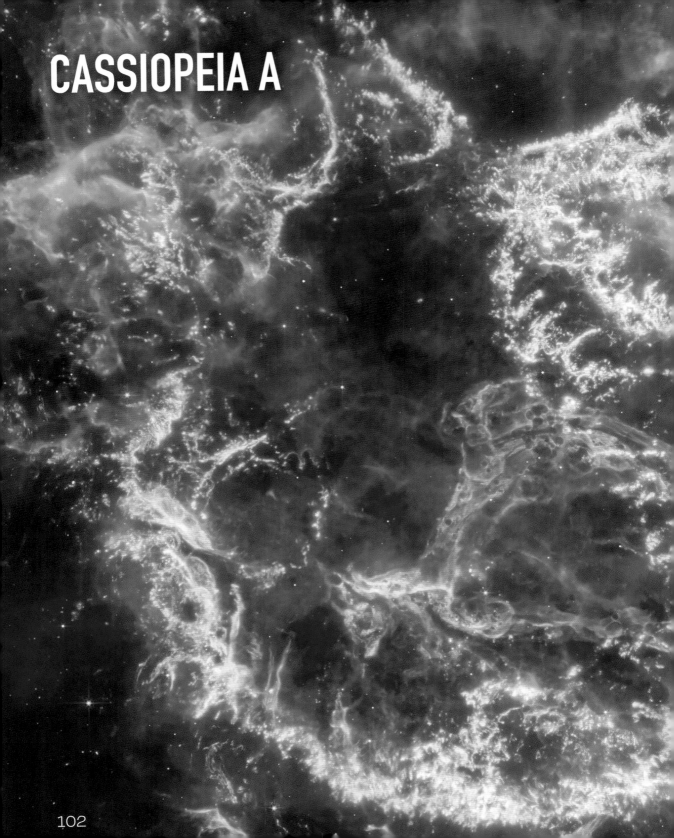

CASSIOPEIA A

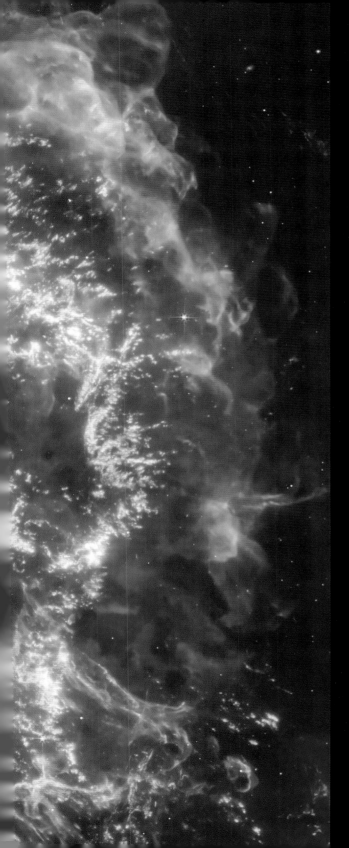

■	F560W
■	F770W
■	F1000W
■	F1130W
■	F1280W
■	F1800W
■	F2100W
■	F2550W

Eleven thousand light-years away, Cassiopeia A (Cas A) stands as one of the most famous examples of a supernova remnant. Spanning 10 light-years, this nebula portrays the chaotic aftermath of a star meeting a violent end. Combined from data taken with eight filters, MIRI's image of Cas A is resplendent with fine details.

Warm dust in the surrounding space glows orange as material from the destroyed star crashes into it. The material itself appears bright pink. Knots and lumps within the expanding cloud look like a burst of liquid caught in a freeze frame.

The complexity of Cas A poses a challenge to astronomers, and an opportunity to investigate the fleeting processes that occur during a supernova.

HUBBLE ULTRA DEEP FIELD – WEBB'S VIEW

The Hubble Ultra Deep Field is one of the most famous regions of the sky, brought to light by the Hubble Space Telescope. Hubble revealed that an apparently blank patch of sky is filled with tens of thousands of galaxies. The original image took 11.3 days to fully expose with Hubble's awesome optical system, but Webb was able to achieve a similar level of sensitivity in just 20 hours. Almost every patch of light in these two panels is a whole galaxy, made up of billions of individual stars. We are looking back through aeons of time.

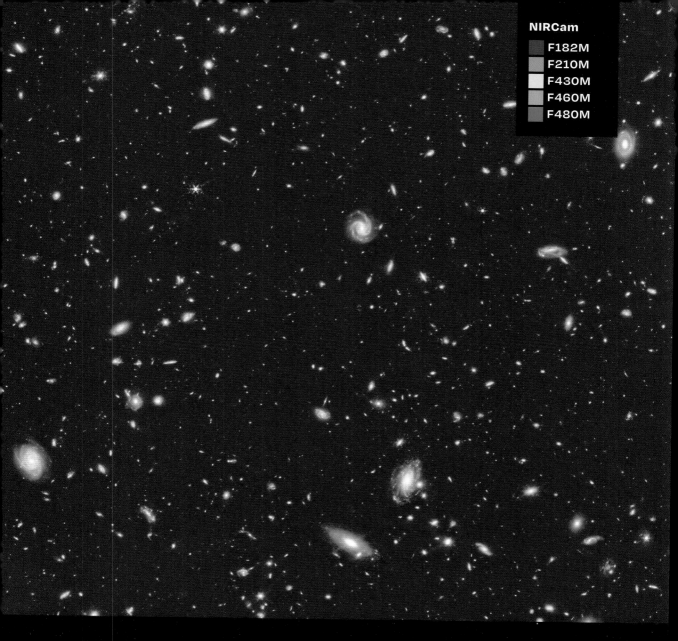

NIRCam

F182M
F210M
F430M
F460M
F480M

Hubble's Ultra Deep Field was an unprecedented undertaking, which
has since inspired subsequent missions to push the boundaries of
how much can be captured in a single image. As a result, Webb will
generate numerous deep field images to help astronomers better
understand the Universe on a grand scale.

NGC 7496

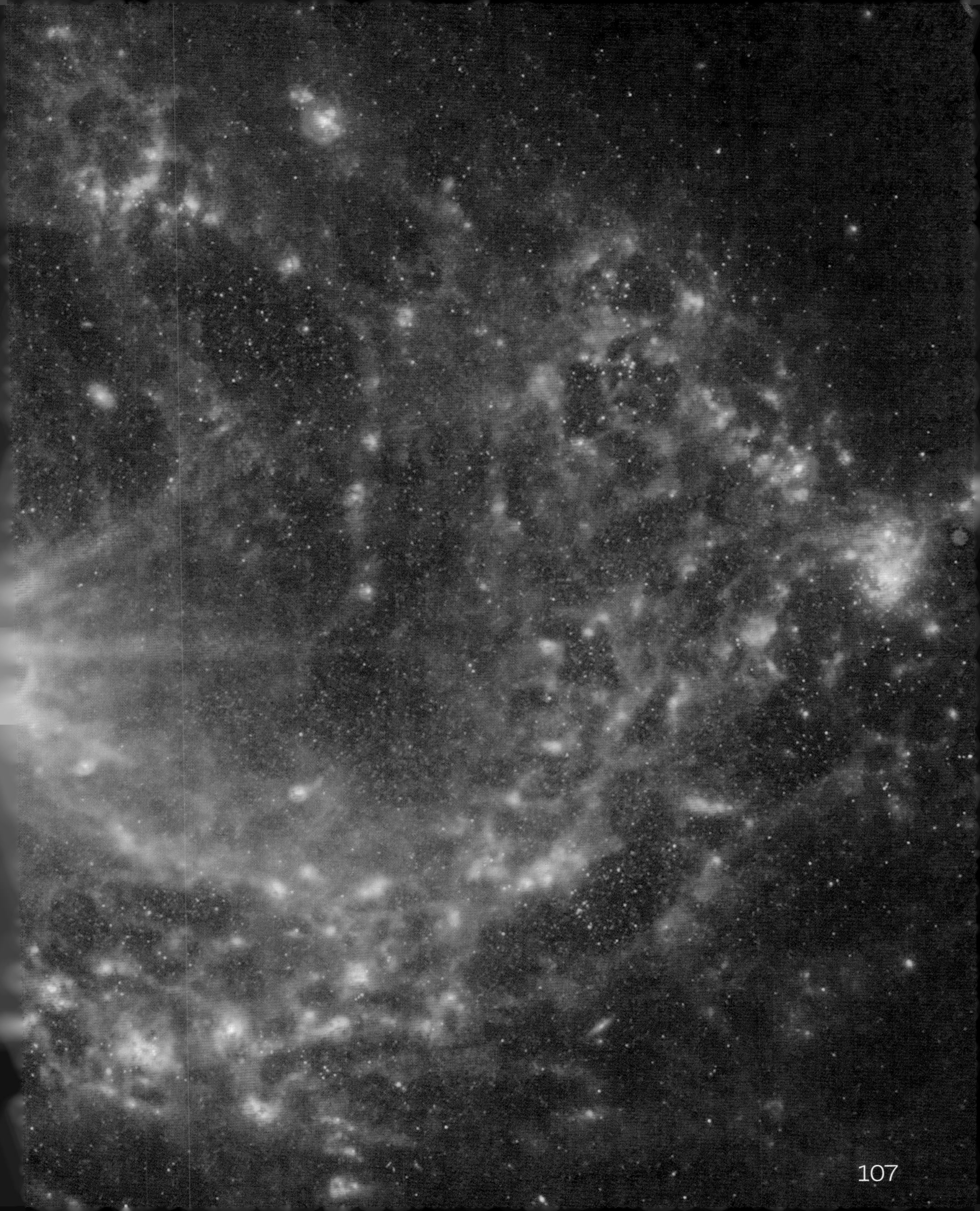

ARP 220

What is happening in this image? It looks like a star shrouded in mottled gas. In fact, Arp 220 is much, much larger. We are watching two spiral galaxies coming together, to ultimately form a single, more massive galaxy at a distance of 250 million light-years from us. Behind them, myriad remote 'island universes' shine in the darkness.

The pair began their interaction about 700 milllion years ago when they first collided, triggering widespread starburst. As a result, about 200 star clusters, brimming with powerful stars, have formed in a compact region roughly 5,000 light-years wide.

Webb's sharp infrared eyes have delineated the cores of the original galaxies, separated by 1,200 light-years. Each is surrounded by an intensely bright ring where active star formation is still underway. The starburst is much brighter than the entire Milky Way Galaxy, casting diffraction spikes onto Webb's detectors.

THE DUSTY DEBRIS DISC AROUND FOMALHAUT

Fomalhaut is a well-studied star, long known to be encircled by a large disc of dusty debris. MIRI has resolved three belts, the largest of which is up to 23 billion kilometres from the star itself. The two inner belts were revealed for the first time by Webb.

Astronomers think that the belts are sculpted by gravitational forces from objects within the disc – most likely unseen exoplanets. Could a system of worlds be hiding just 25 light-years away?

COMET 238P/READ

NIRCam

F200W

In September 2022, astronomers took the opportunity to point Webb's mirror towards Comet 238P/Read, a 'main belt' comet discovered in 2005. 238P/Read has the characteristic features of any other comet: a coma, or hazy atmosphere, and a long tail made of ices that have vapourised from the surface in the warmth of the Sun, but Webb's instruments detected that the comet lacked any significant carbon dioxide gas in its coma. Rather, its extended features are abundant in water vapour.

Nevertheless, Webb found that 238P/Read generates its coma and tail through the same mechanisms as a comet from beyond the orbit of Neptune, advancing our understanding of main belt comets.

EXTRAGALACTIC EXPANSE

NIRCam

- F090W
- F115W
- F150W
- F200W
- F277W
- F335M
- F356W
- F410M
- F444W

Following in the footsteps of the Hubble Space Telescope, Webb captured this NIRCam image as part of the JWST Advanced Deep Extragalactic Survey (JADES) programme. In total, over 45,000 galaxies are represented here, with several hundred seen at a time when the Universe was less than 600 million years old.

This period of cosmic history has eluded scrutiny, requiring extreme infrared sensitivity to be properly observed. But thanks to Webb, astronomers can identify galaxies that inhabited the so-called Era of Reionisation, during which the expanding Universe became transparent, as electrically neutral atoms separated into ions.

By analysing these galaxies, astronomers will learn more about this formative chapter in the story of the Universe.

IC 5332

C 5332 is a Spiral Galaxy about 30 million light-years away. To find out more about this image, see page 48.

QUASAR
J0100+2802

In deep field images like this one, which contains over 20,000 galaxies, stars within the Milky Way cast diffraction spikes through Webb's optical system. But not every object with spikes in this image is a star. NIRCam has been deliberately centred on a bright point source, which looks like a pink star. In fact, this is a quasar, or quasi-stellar object. Like most of the objects in the scene, it is extremely far away. The properties of quasars were once a mystery, but today we know they are super massive black holes that are highly active, feeding on the material around them and blazing with energy.

In this image, galaxies closer to us are being analysed as they are 'backlit' by the quasar. Using this method, astronomers have been able to determine that during the Era of Reionisation, it was active galaxies that caused the Universe to become transparent by ionising the gas around them.

ORION BAR
(NIRCAM)

NIRCam

- F140M
- F162M
- F164N
- F182M
- F210M
- F187N
- F277W
- F212N
- F300M
- F335M
- F323N
- F470N
- F480M

I n a dreamlike, painterly scene, NIRCam allow us to gaze deeply into the heart of the Orion Nebula, to admire the interplay of gas, stellar wind and ultraviolet radiation. Stargazers have long admired the Trapezium Cluster, which is out of the frame to the upper left. Its stars are sculpting this feature, called the Orion Bar, through a process of erosion.

A young Solar System to be, d203-506, comprises an infant star and protoplanetary disc. Webb detected methenium ($CH+_3$), also known as methyl cation, within this disc. It marks the first such detection outside our own Solar System. The existence of carbon compounds near such strong ultraviolet sources is compelling, and suggests the building blocks of life may be more commonly found throughout the Galaxy than previously expected.

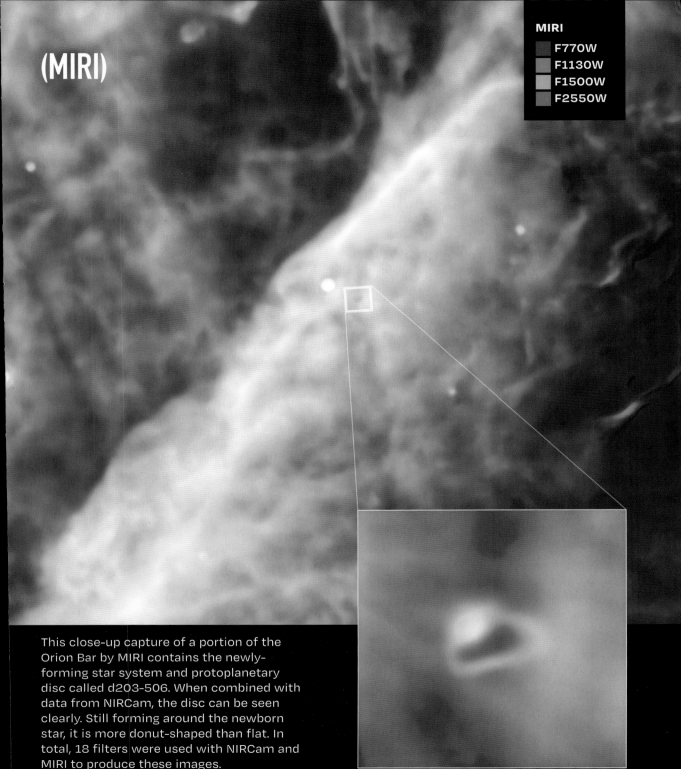

(MIRI)

MIRI

▢ F770W
▢ F1130W
▢ F1500W
▢ F2550W

This close-up capture of a portion of the Orion Bar by MIRI contains the newly-forming star system and protoplanetary disc called d203-506. When combined with data from NIRCam, the disc can be seen clearly. Still forming around the newborn star, it is more donut-shaped than flat. In total, 18 filters were used with NIRCam and MIRI to produce these images.

ASPIRE COSMIC FILAMENT

NIRCam
F115W
F200W
F356W

Though this field appears to be littered with galaxies scattered throughout the cosmos, astronomers have identified that ten of the sources in this image form a three-million-light-year-long filament, which may be in the process of developing into a large cluster. The filament galaxies appear as they were just 830 million years after the big bang, during the Era of Reionisation.

One of the cornerstones of the Webb mission is to help astronomers further understand the early evolution of the Universe, and its extraordinary sensitivity has proved to be invaluable in advancing this cosmological investigation.

NGC 5068

NGC 5068 is a Barred Spiral Galaxy about 20 million light-years away. To find out more about this image, see page 48.

SATURN

Dione

Enceladus

Tethys

S aturn, the jewel of the Solar System, is a favourite for amateur astronomers to find with their telescopes. Ordinarily, the orb and famous rings of this beautiful planet shine with comparable brilliance, but in this image it looks like someone has turned the lights out in Saturn's atmosphere. That's because methane in the Saturnian atmosphere absorbs near-infrared radiation from the Sun. In contrast, the icy rings stay bright in these wavelengths.

NIRCam captured this image as part of a test of Webb's capabilities to detect faint moons around Saturn and its rings. At the time of writing, it holds the record for the most known moons anywhere in the Solar System, with at least 146.

SATURN'S MYSTERIOUS MOON: TITAN

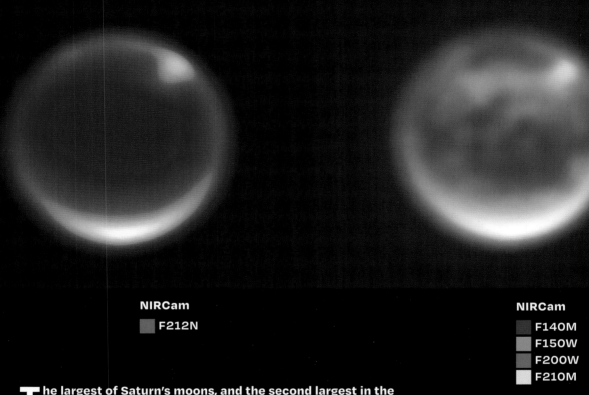

NIRCam
- F212N

NIRCam
- F140M
- F150W
- F200W
- F210M

The largest of Saturn's moons, and the second largest in the Solar System, Titan is an extraordinary world in its own right. It is the only planetary satellite to boast its own atmosphere, which is explored in these images from NIRCam. On the left, a single channel captured using the 2.12-micron filter shows Titan's lower atmosphere. Bright spots indicate the presence of clouds, which are also seen in the three-channel image, detailing large surface features. In visible wavelengths, Titan's atmosphere is opaque, all but hiding its alien terrain. Peering through the thick haze, NIRCam shows us Kraken Mare – a large body of liquid ethane and methane in the northern polar region – and the equatorial 'sand sea' of dunes and plains known as Belet. Adiri is a particularly bright region, seen here on the right size of the image.

DUSTY SUPERNOVAE IN THE FIREWORKS GALAXY

SN 2004et

MIRI observed the remnants of not one, but two Type II supernovae in NGC 6946, the Fireworks Galaxy, still glowing long after the stars exploded. What once were stars are now expanding, dusty clouds, clearly seen by MIRI despite being 22 million light-years away.

Webb has resolved both cooler dust (shown in red) and hotter dust (with bluer hues). The observations of abundant dust provide evidence for the theory that supernovae were efficient dust-generators, which enriched their parent galaxies as the early Universe evolved. Researchers calculated that SN 2004et, which is 13 years older and cooler than its sibling across the galaxy, has released a quantity of dust more than 5,000 times as massive as Earth.

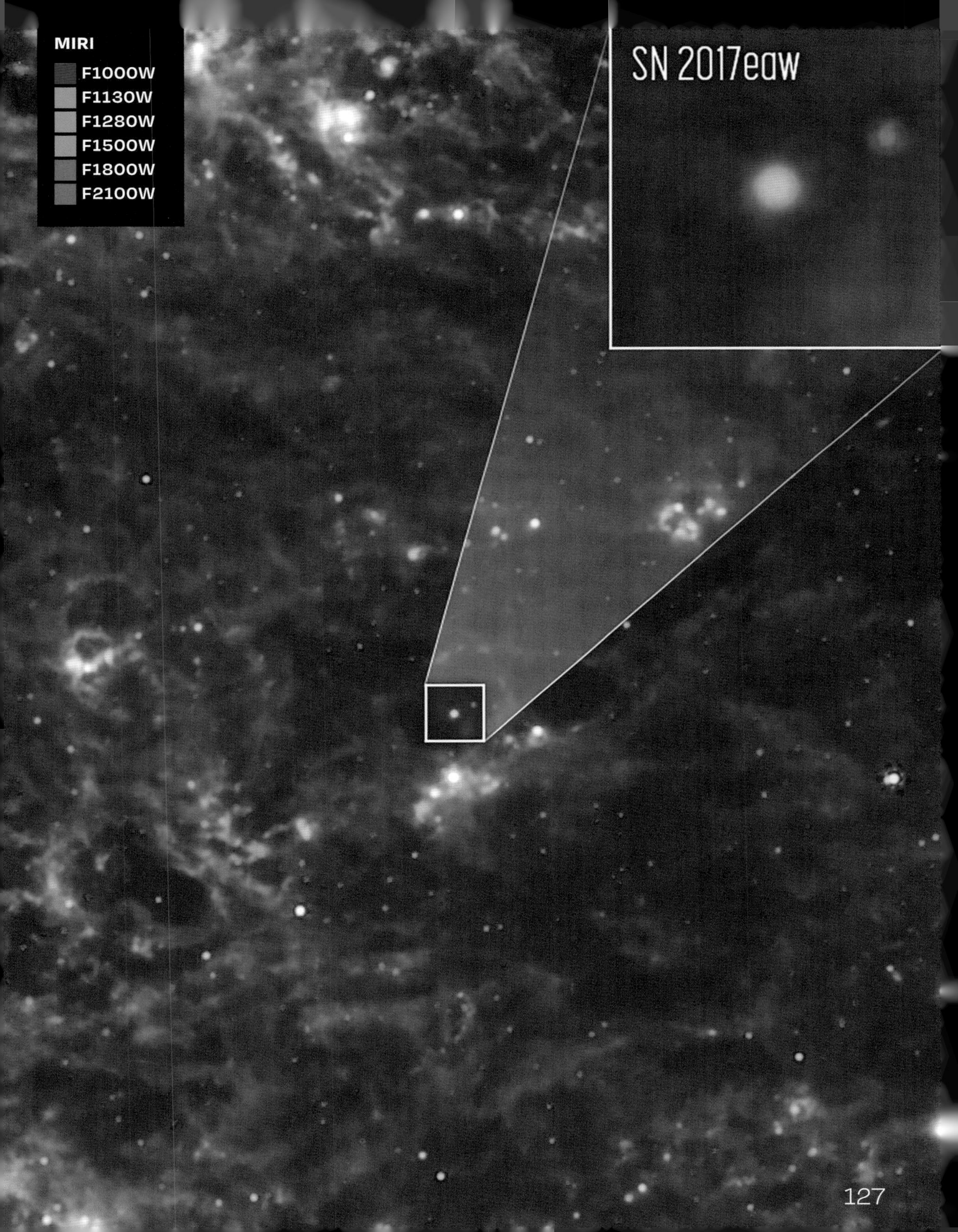

MIRI

F1000W
F1130W
F1280W
F1500W
F1800W
F2100W

SN 2017eaw

CEERS SURVEY

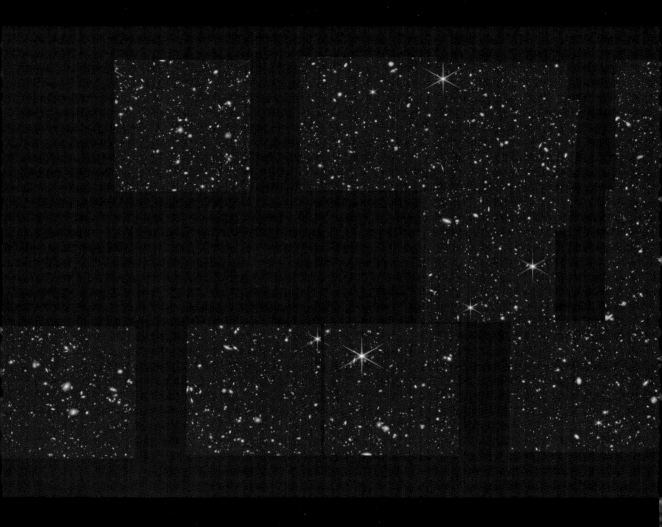

An ocean of galaxies fill this cosmic vista, assembled from images captured by NIRCam. Many show a magnificent range of details, from sweeping spiral arms to sprinkles of star clusters. Just a few stars in our own Milky Way pollute the scene, casting bright blue diffraction spikes onto the detector.

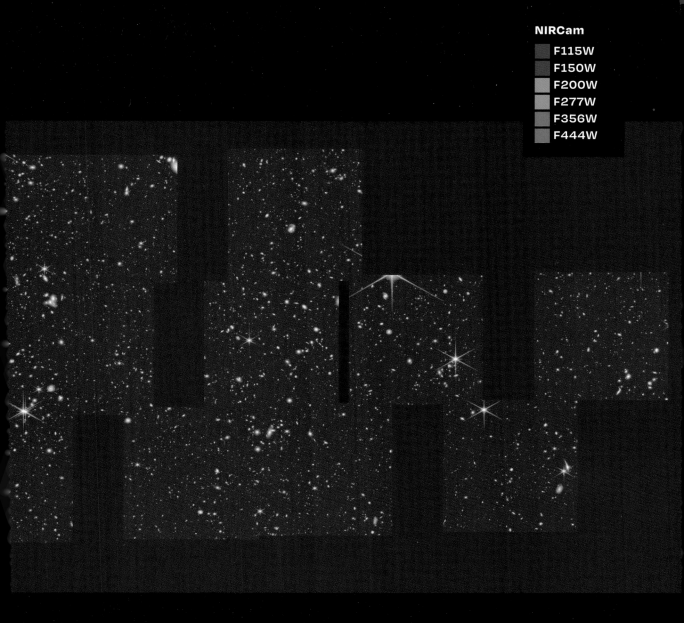

This partially assembled image is part of the Cosmic Evolution Early Release Science (CEERS) Survey, and it features a staggering number of galaxies – more than 100,000 in total. Some of them, appearing as red dots, are seen as they were over 13 billion years ago, during the first age of the Universe. Many mysteries remain about the lives of galaxies during this formative period, but Webb has already revealed more extant galaxies during this time than astronomers expected to see. For all its investigative capabilities, Webb is also generating new questions that challenge our understanding.

NGC 4535

NGC 4535 is a Barred Spiral Galaxy about 50 million light-years away. To find out more about this image, see page 48.

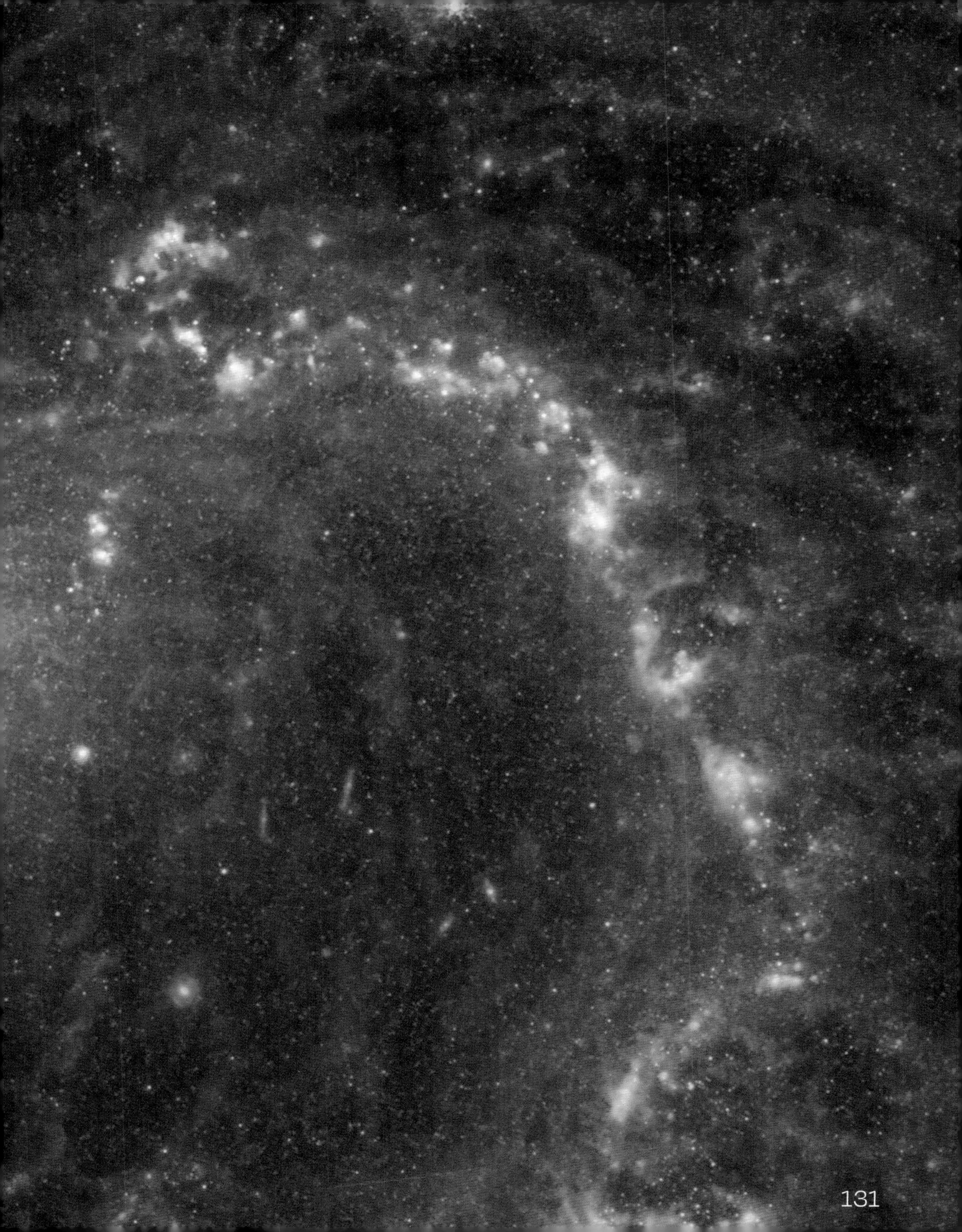

RHO OPHIUCHI

For the first anniversary of its scientific operations, Webb's NIRCam was pointed at the stunning Rho Ophiuchi cloud complex – the nearest star-forming region in the Galaxy. Compared with other star nurseries, it's a calm and quiet place, but in the infrared, it takes on a tumultuous character. Young stars emit jets that plough their way through the interstellar cloud, energising molecular hydrogen and causing it to glow red in the image.

Some of the stars are surrounded by discs of material, from which planets may form. The star S1 illuminates the cavity in the lower half of the image. The cream-coloured gas in this region contains polycyclic aromatic hydrocarbons, which are thought to be ideal starting material for the building blocks of life.

This scene spans a width of approximately one light-year, and the nebula is home to about 50 stars.

H 46/47

Two young stars, still forming, are locked in a dance. They are tightly bound by their mutual gravity, buried within a dense cloud of gas and dust, from which they are feeding and growing. They are so close together that, in this image, they appear as a single bright source creating pink and orange diffraction spikes. Together, they form a Herbig-Haro object, HH 46/47. The stars are eating at such a high rate that they regularly become unstable and eject material, forming the two prominent lobes of gas stretching out to the sides of the image.

The left-hand lobe has interacted more prominently with the surrounding nebula, shown in blue here, causing it to widen and glow brightly. These stars are just a few thousand years old. For millennia to come, their eruptions will disrupt the material around them, regulating how quickly they feed as they settle on a more steady diet. In a few million years, they will be fully formed.

EL GORDO

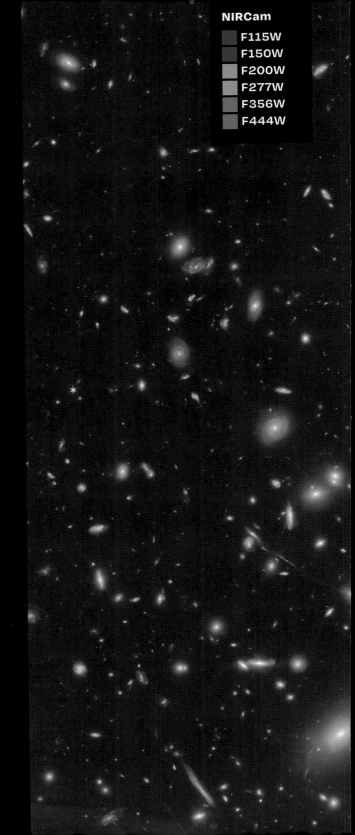

NIRCam
F115W
F150W
F200W
F277W
F356W
F444W

Look closely and you'll see some very strange galaxies in this deep field NIRCam image of the galaxy cluster called El Gordo. In fact, these are ordinary galaxies whose images have been distorted by the presence of a massive gravitational lens. The mass of El Gordo warps space on an intergalactic scale, which in turn alters the path of light reaching us from behind the cluster. Astronomers theorise that this massive lens, which does not absorb light itself, is composed of dark matter.
The image contains just a handful of stars from our own Galaxy, which are conspicuous by their diffraction spikes.

NGC 4321

N GC 4535, also known as M100, is a Spiral Galaxy about 55 million light-years away. To find out more about this image, see page 48.

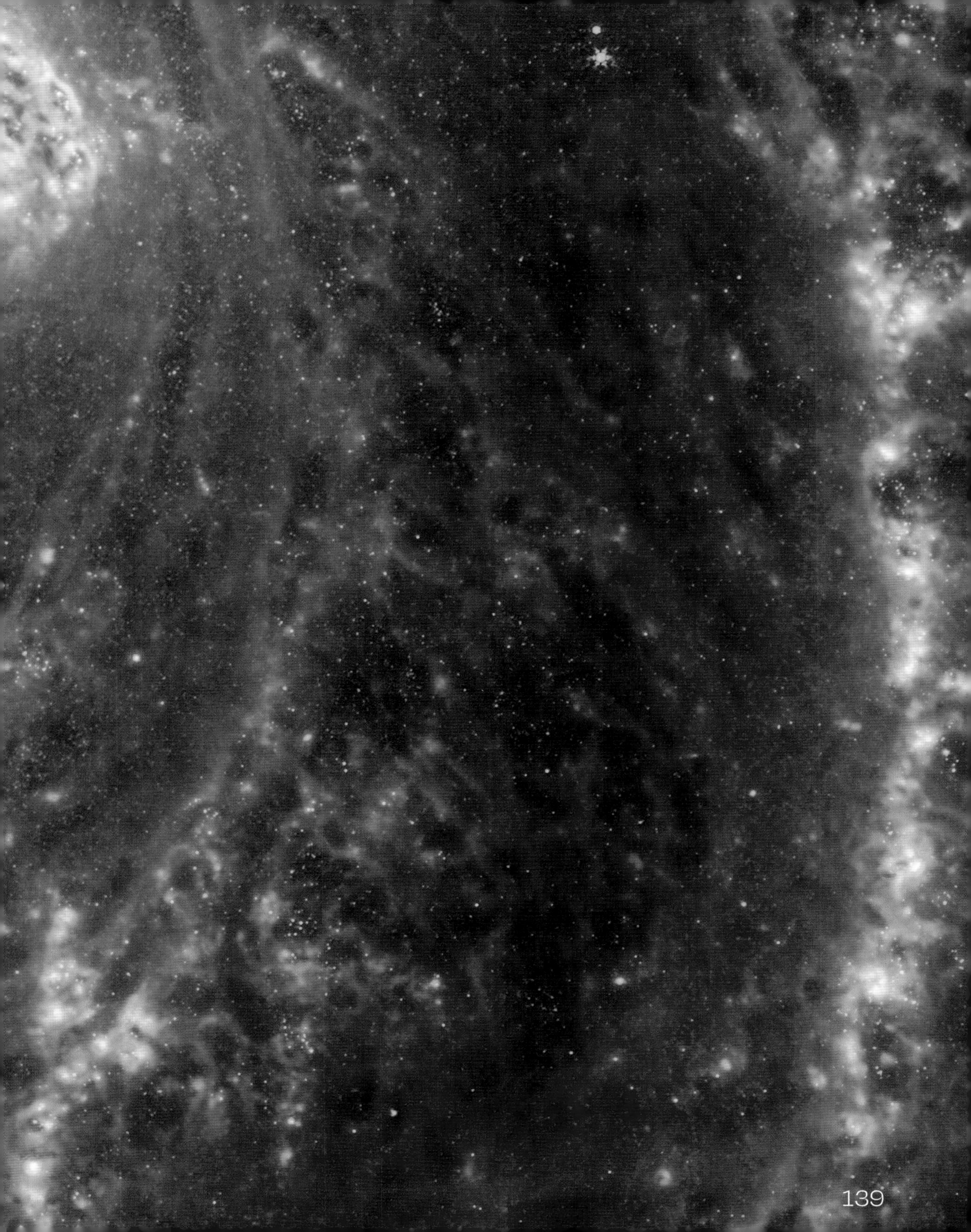

SUNRISE ARC

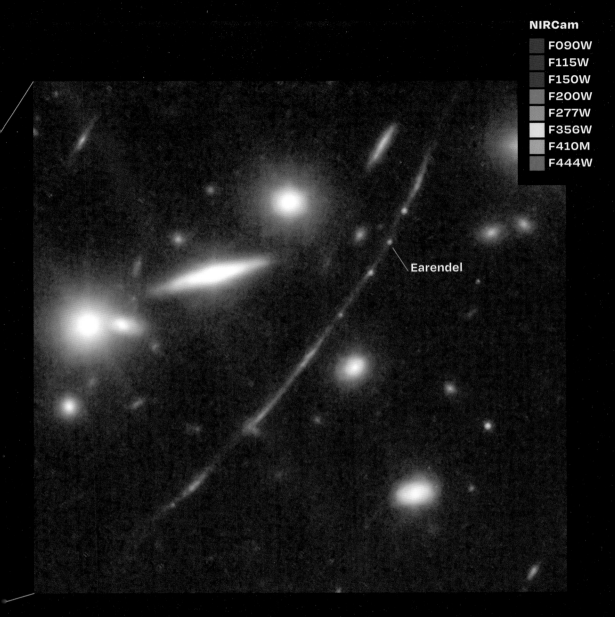

NIRCam

■	F090W
■	F115W
■	F150W
■	F200W
■	F277W
■	F356W
■	F410M
■	F444W

Earendel

The galaxy cluster WHL0137-08 forms a notable gravitational lens, which creates the most strongly magnified image of a galaxy from the first billion years of cosmic history to have been discovered. It can be seen as a thin orange line to the lower right of the cluster itself. The gravitational magnification is so great that a single, powerful star has been resolved – the most distant star ever detected. It has been named Earendel, after the half-elven seafarer from J. R. Tolkien's The Silmarillion, Eärendil, who wore on his head a brilliant jewel called the Morning Star. The Morning Star in the Sunrise Arc shines new light on the study of the early Universe.

THE RING NEBULA
(NIRCAM)

The Ring Nebula is a popular target for amateur astronomers. Visible even in small telescopes, it appears as a perfect 'smoke ring' in the sky. Webb's view is considerably grander. Here we see the famous planetary nebula through the eyes of both NIRCam and MIRI. A dying star roughly 2,500 light-years away is in the process of shrugging off its outer layers into the surrounding expanse.

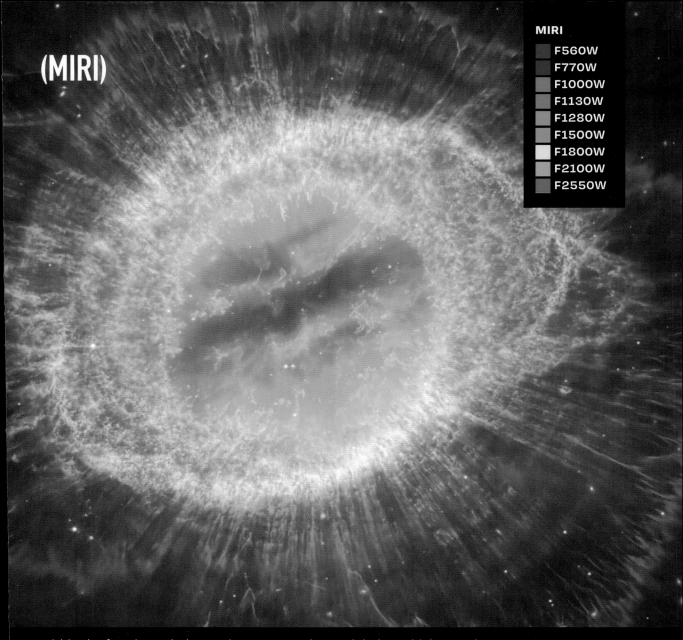

(MIRI)

Within the forming nebula are about 20,000 dense globules, which contain abundant molecular hydrogen. What seems to be a pool at the centre, blue to NIRCam and red to MIRI, is a bubble of hot gas. The intricate outer layers, including the ten concentric rings, are cooler. These structures were probably formed as a result of the dying star interacting with a close companion, with the two being separated by about the same distance as Earth and Pluto. These exquisite portraits of the Ring Nebula will enable astronomers to unravel more details about the end-of-life processes for stars like the Sun.

NGC 5584
(WEBB AND HUBBLE)

Hubble and Webb form an astronomical dream team for the creation of this galactic portrait. Seventy-two million light-years away, spiral galaxy NGC 5584 is pictured in a combination of visible and infrared light, with NIRCam data shown in red. This channel isolates members from a particular classification of star, known as Cepheid Variables, from their neighbouring stars.

Webb goes beyond the resolution limit of Hubble, to provide more accurate observations of these stars, which are used by astronomers to measure cosmic distances. The observations made for this image confirmed that Hubble has been on the right track for decades, providing astronomers with more confidence in their calculations related to the expanding Universe.

NGC 4303

NGC 4303 is a Starburst Galaxy about 55 million light-years away. To find out more about this image, see page 48.

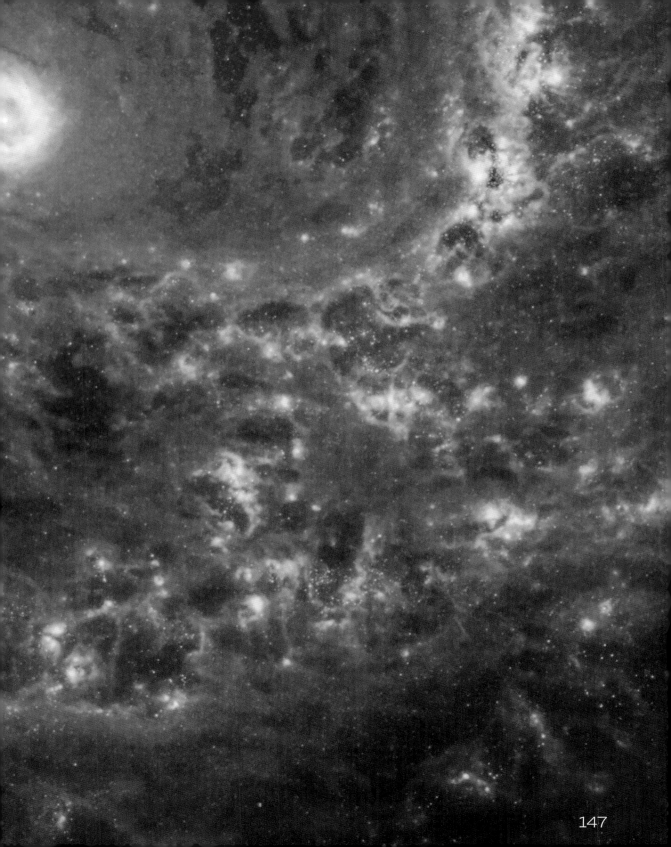

HH 211

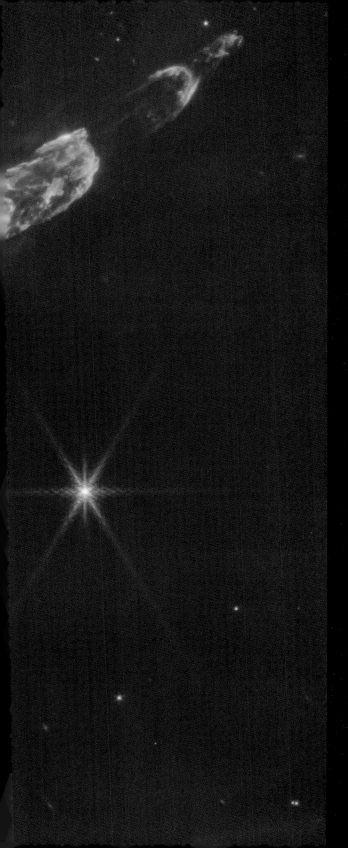

NIRCam

■	F210M
■	F212N
■	F335M
■	F466N
■	F460M
■	F470N

A Herbig-Haro object 1,000 light-years away puts on a dramatic show for NIRCam. A young star, hidden even from Webb by a very dense cloud, is spewing out gas in two directions to form the long pink lobes. The jets are carried on strong stellar winds as they plough through the larger nebula. High-speed collisions heat the gas and dust, causing the shock-fronts of the expanding lobes to glow.

The image shows extraordinary detail and texture, offering astronomers an opportunity to trace the history and chemistry of the clouds, which contain molecular hydrogen, carbon monoxide and silicon monoxide.

JUPITER

NIRCam

F164N
F212N
F360M

As with all Webb images, this multi-channel composite of Jupiter has a striking and unusual palette. While it may not look like the familiar, visible, light images of Jupiter, with their warm hues – it does contain a wealth of information about Jovian weather and space weather. Bright spots indicate high-altitude features in Jupiter's cloud tops, whereas the red glows over the poles are emissions from powerful Jovian auroras.

Between the darker belts, which have relatively little cloud cover, is the bright equatorial region. Webb detected for the first time a powerful, narrow jet stream above the main cloud decks, travelling at 320 miles per hour.

EUROPA

NIRCam

- F070W
- F140M
- F212N
- F250M
- F335M
- F460M

Meanwhile, NIRCam has also been employed to observe the mysterious and fascinating ice world Europa, one of Jupiter's four large Galilean Moons. Astronomers have long been fascinated by its subsurface salt-water ocean, which will be scrutinised intensely by both the NASA Europa Clipper and ESA JUICE missions, as Webb is reaching the end of its mission. The data from the telescope indicates the presence of carbon dioxide ice on the surface, particularly in the brighter region, which probably originated from the ocean underneath. Could life be lurking there?

NGC 346

MIRI

- ■ F770W
- ■ F1000W
- ■ F1130W
- □ F1500W
- ■ F2100W

In a bright corner of the Large Magellanic Cloud, a dwarf satellite galaxy of the Milky Way, lies the young star cluster NGC 346. It is enshrouded by a complex and beautiful nebula, shown here in blue, which is rich in silicates and polycyclic aromatic hydrocarbons. The nebula is highly active with numerous stellar nurseries. Protostars embedded within the dust appear as bright spots and filaments. The nebula is approximately 200 light-years wide, and roughly 200,000 light-years away from us.

NGC 4254

N GC 4254, also known as M99, is a Spiral Galaxy about 50 million light-years away. To find out more about this image, see page 48.

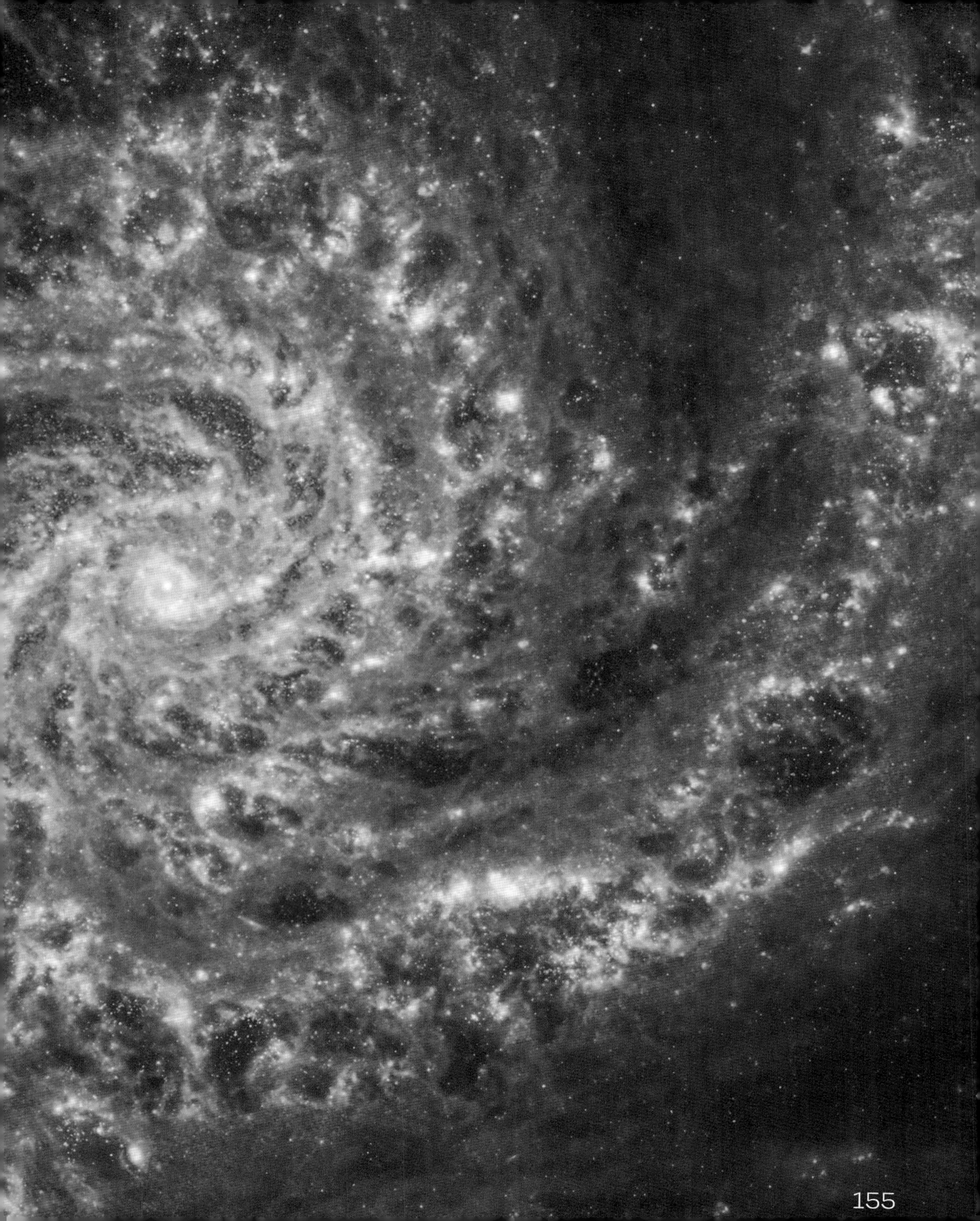

KILONOVA AND
HOST GALAXY

This image pushes the boundaries of what can be captured in a single point of light. The red dot in the upper left tells a remarkable story. Over a billion years ago, a pair of extremely dense objects – two neutron stars, or one neutron star and one black hole – were ejected from their host galaxy (right). After travelling about 120,000 light-years – a distance equivalent to the diameter of the Milky Way – over a period of millions of years, the pair finally spiralled in and merged, resulting in a cataclysmic kilonova. As the name implies, the process is likened to a supernova, but it is much more powerful. In this case, the kilonova produced an observable Gamma Ray Burst (GRB) that was a thousand times more powerful than is typical for such an event. Webb is perfectly suited to studying extreme astrophysical events at tremendous cosmic distances.

THE CRAB NEBULA

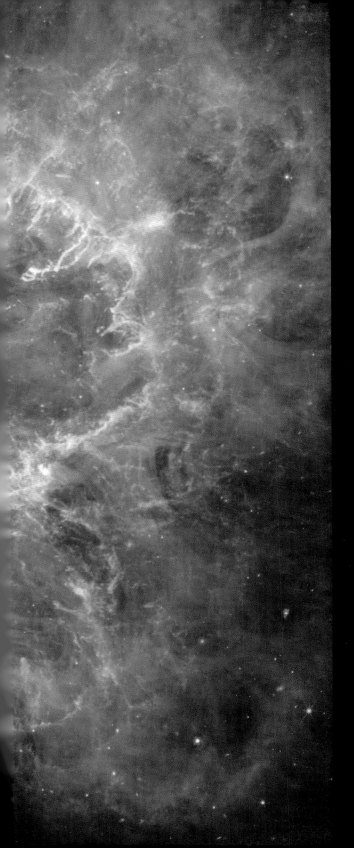

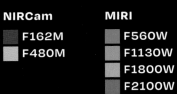
The Crab Nebula is the majestic remnant of a supernova that was seen by sky watchers in 1054 and recorded by Chinese astronomers as a 'guest star'. Rusty orange filaments and cliffs of doubly ionised sulphur create a vibrant contrast to the hotter gas flooding the nebula's core regions. Bluer filaments contain ionised iron, scattered during the massive star's violent end, whereas the pallid, smoke-like whisps reveal synchrotronic radiation, in which particles are accelerated to enormous speed as they spiral around magnetic field lines.

What is the source of all this magnetic energy? The answer lies at the very heart of the Crab Nebula, where a spinning neutron star – also called a pulsar – appears as a brilliant white point. You can find it surrounded by the most concentrated magnetic flux lines.

MACS 0416
(WEBB AND HUBBLE)

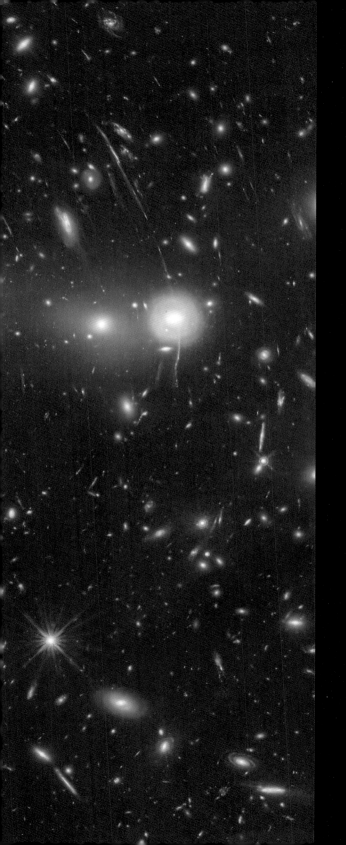

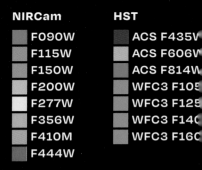

NIRCam	HST
■ F090W	■ ACS F435V
■ F115W	■ ACS F606V
■ F150W	■ ACS F814W
■ F200W	■ WFC3 F105
■ F277W	■ WFC3 F125
■ F356W	■ WFC3 F14C
■ F410M	■ WFC3 F16C
■ F444W	

Fifteen filters were used in this impressive composite image from Webb's NIRCam and Hubble's Wide Fie Camera 3 (WFC3) and Advanced Came Surveys (ACS), which combines light the visible and infrared spectra. The re is an astonishingly colourful, deep fie image of the galaxy cluster MACS 041

The cluster, about 4.3 billion light-yea away, creates a gargantuan gravitatior lens, distorting and magnifying a dizzy array of background galaxies. You cou say the image involves not two, but th instruments, as gravitational lenses lil this one can magnify images of very d supernovae, or individual stars, in the remote galaxies. This image was comp as part of a survey to find them. It's vi and scientific quality is a testament to the enormous power of precise, multiwavelength space-based astron

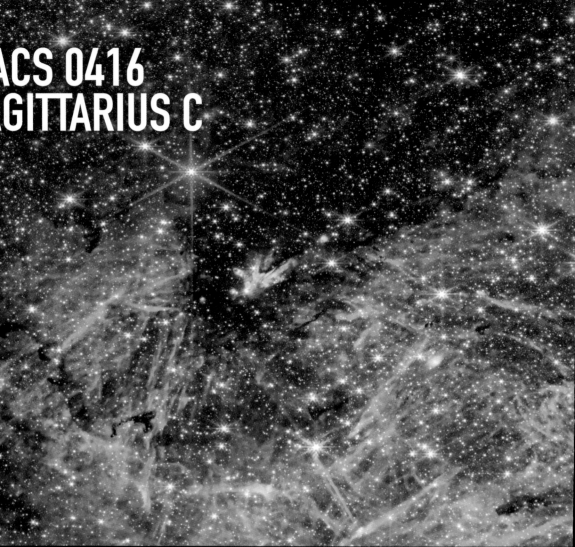

ACS 0416
GITTARIUS C

e half a million stars populate this vista, which depicts a small
rtion of the Milky Way's dense central region, spanning just 50
ears. The cyan cloud, which encompasses half the width of the image,
e of ionised hydrogen. Above it, what appears to be a relatively empty
is actually a cloud too dense to be peered into.

NIRCam

F162M

F360M

F405N

F470N

The cyan cloud is wrapped around it. This 'infrared-dark cloud', impenetrable even to NIRCam, could be a future birthplace for massive stars. At least a portion of it, near the bottom and shown in pink, is already active. Here, a cluster of protostars are beginning to emerge from a dense cocoon of gas. Webb's observations of the Milky Way's core are a treasure trove for astronomers, full of features that have never before been revealed

NGC 3627

NGC 3627 is a Spiral Galaxy about 36 million light-years away. To find out more about this image, see page 48.

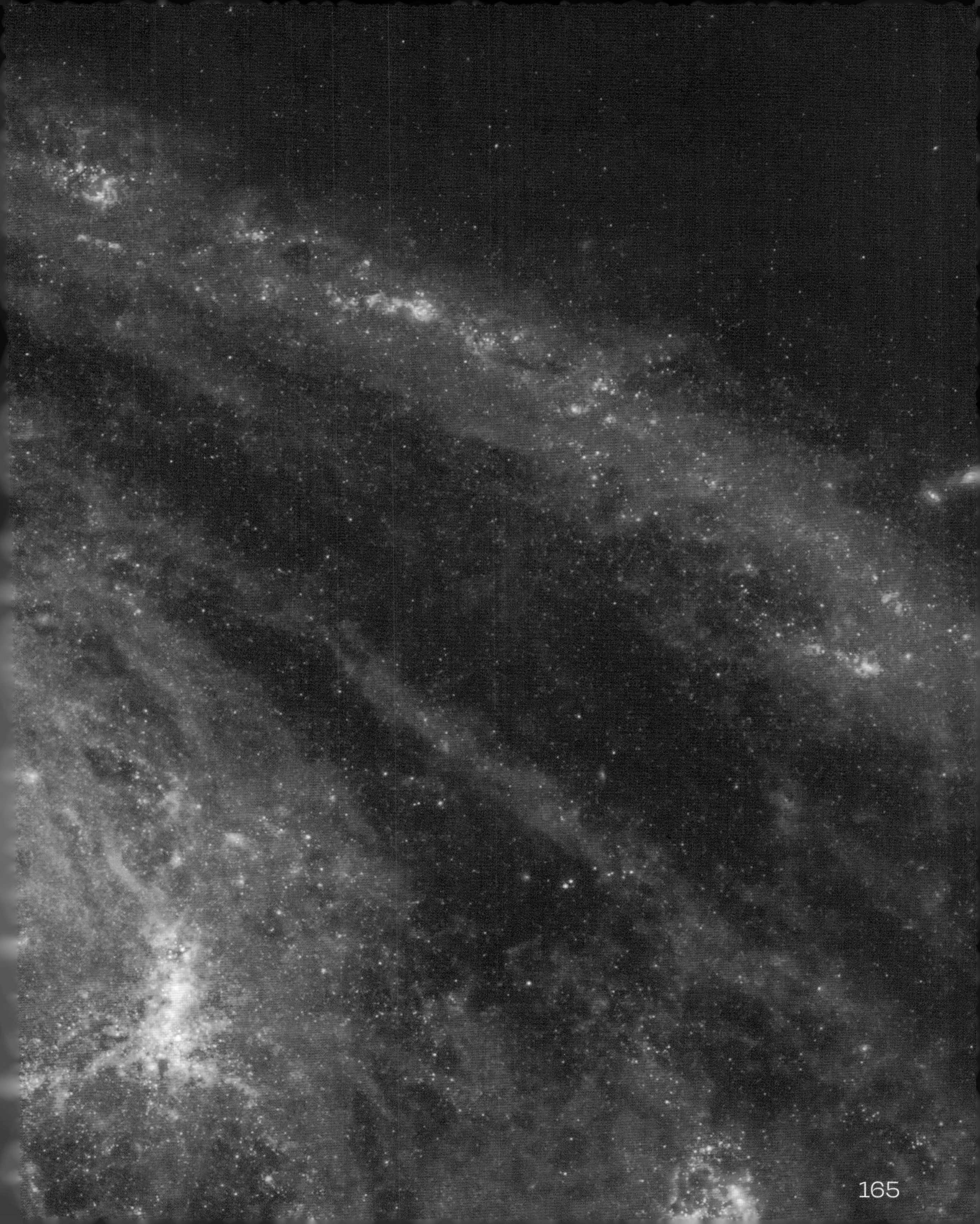

CASSIOPEIA A
(NIRCAM)

The Cassiopeia A supernova remnant comes to life in this razor-sharp image, perfectly illustrating the sheer chaos that results from a massive star exploding. An expanding shell of material is crashing through a larger gaseous cloud, which was released more gently over a long period before the supernova occurred. Where the two meet, hot shock-fronts appear incandescent. Finely resolved globules of pink and orange contain sulphur, oxygen, argon, neon and other elements cast out by the supernova.

What remains of the shattered star's core, a neutron star, produces strong magnetic fields that permeate the remnant. Synchrotron radiation, caused by accelerated particles riding the magnetic flux, appears as a ghost-white veil around the nebula.

IC 348

C 348 is a star cluster in the local region of the
Milky Way, about 1,000 light-years away from us.
Here we see just its central region, where the light
of brilliant stars is illuminating adjacent, fluffy
clouds of gas and dust, forming a reflection nebula.
Astronomers captured this image as part of a search
for brown dwarfs – warm substellar objects – within
the cluster, and found three that are less than eight
times the mass of Jupiter. The brightest star at the
centre of the image is not, in fact, a single star, but
two stars tightly bound in a binary system. Their
strong stellar winds may play a role in sculpting the
painterly nebula's loop-like structure.

NEBULOSITY IN THE PLEIADES

The Pleiades is a familiar sight in our night sky, as a bright and splendid star cluster in the constellation of Taurus. As a collection of hot stars, they shine distinctly blue in the visible part of the spectrum, as shown in the lower image. They cast their light onto surrounding gas clouds, forming a reflection nebula. This NIRCam image captures a small portion of the Pleiades reflection nebula near the star Merope, whose enormous diffraction spikes beam across the field. The nebula appears yellow and magenta-toned, with a texture like a fast-moving stream. The exposure is so deep that numerous background galaxies can be seen behind it. This image was captured as part of a programme to look for ambipolar diffusion in the nebula. The process involves neutral hydrogen molecules decoupling from plasma that is bound to the interstellar magnetic field, and is necessary for the gravitational collapse that leads to star formation.

NIRCam
F220W
F277W
F225M

171

NGC 3351

NGC 3351 is a Barred Spiral Galaxy about 33 million light-years away. To find out more about this image, see page 48.

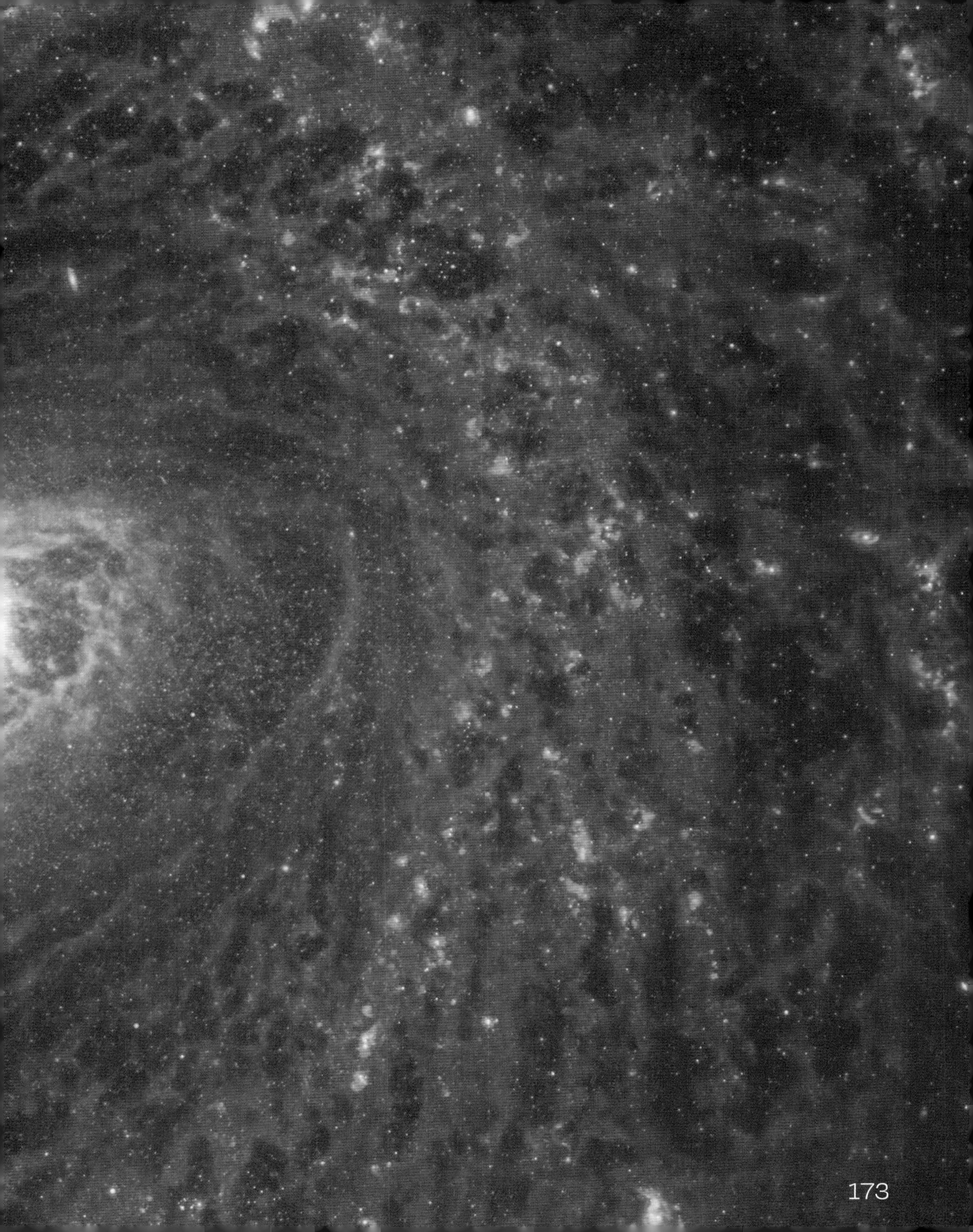

GOODS-NORTH FIELD

This vast expanse of galaxies, known as the GOODS-North field, was captured as a high-resolution mosaic by NIRCam to aid the study of the early Universe. One unassuming point of light is particularly interesting. The circled galaxy, just a dot in the image, is GN-z11. It is one of the youngest galaxies ever observed, appearing as it was just 430 million years after the big bang.

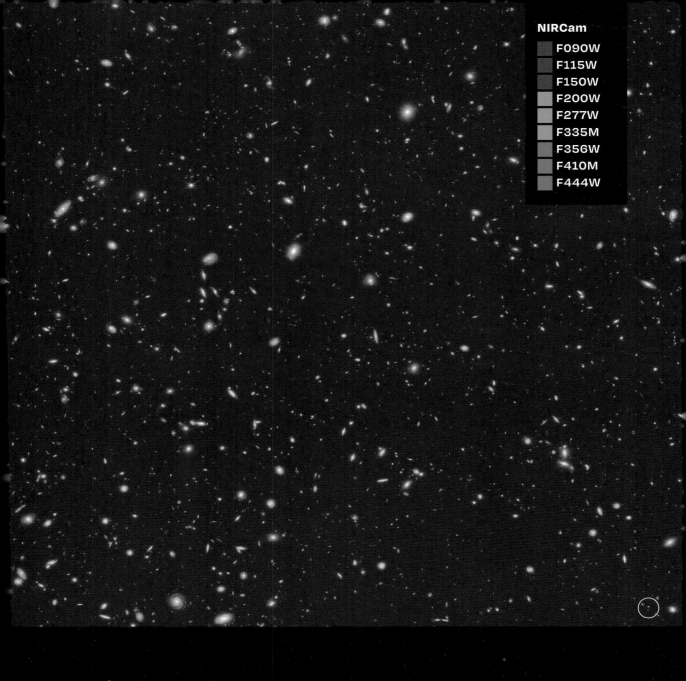

NIRCam

F090W
F115W
F150W
F200W
F277W
F335M
F356W
F410M
F444W

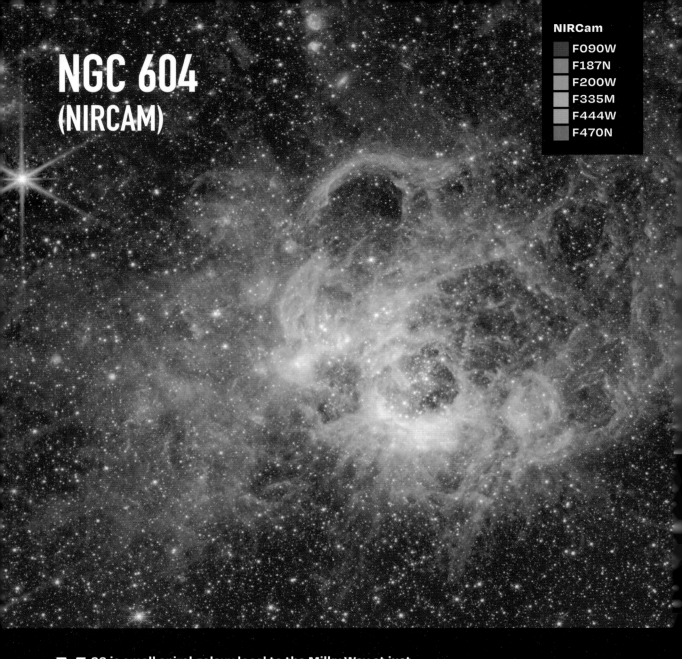

NGC 604
(NIRCAM)

NIRCam

F090W
F187N
F200W
F335M
F444W
F470N

M 33 is small spiral galaxy local to the Milky Way at just 2.73 million light-years away. Here we see an impressive star-forming region embedded within one of its spiral arms. Tendrils of red gas, molecular hydrogen, are stretched out by the powerful stellar winds of young stars, which carve out cavities in the nebula. For millions of years to come, stars will continue to be born from these fertile clouds.

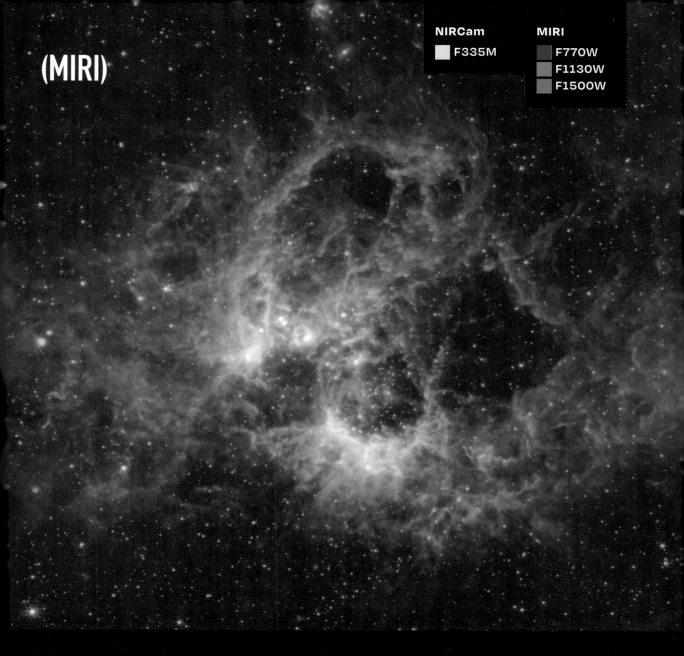

(MIRI)

NGC 604 takes on a ghostly visage in this composite image, which combines brightness data from NIRCam with colour data from MIRI. Over 200 powerful stars reside within the nebula, whose population is set to grow as active star formation continues. While this image shows fewer hot stars than the NIRCam image alone, it does contain a few red supergiants. Such stars are cooler than the Sun, but much larger, sometimes hundreds of times greater in diameter.

NGC 5468
(WEBB AND HUBBLE)

A face-on spiral galaxy gleams in full glory thanks to the combined power of Hubble and Webb. Hubble's WFC3 provides a sharp view of NGC 5468's spiral arms over a distance of some 130 million light-years. Meanwhile, NIRCam's 2.77 micron filter provides the red channel, singling out Cepheid-variable stars. Their distances can be determined by astronomers, and in this case the calculations were corroborated by observations of type Ia (type one-A) supernovae within the galaxy. That the two methods of distance determination agree provides renewed confidence in established measurements of the scale of the Universe.

NGC 2835

NGC 2835 is a Barred Spiral Galaxy about 35 million light-years away. To find out more about this image, see page 48.

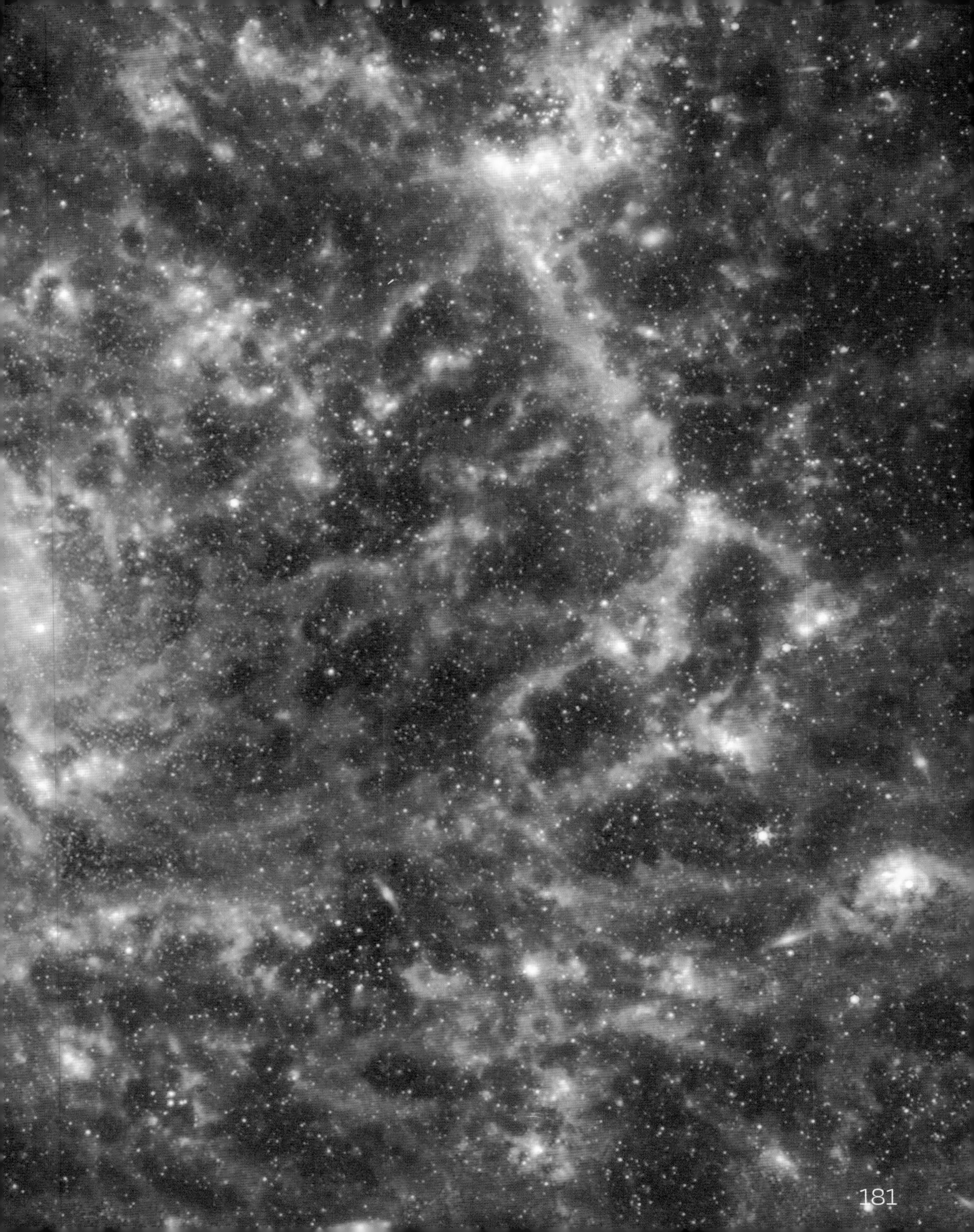

PROTOSTAR IRAS 233854

This is a single-channel image captured by MIRI through the 15-micron filter. It depicts a protostar called IRAS 23385, which is forming within a chemically rich nebula. This observation was made as part of a programme to search for the key ingredients necessary to form potentially habitable exoplanets around protostars, using spectroscopic infrared data. A team of astronomers determined that various organic molecules are locked up inside ices within the interstellar medium, including ethanol and methane. It is possible that this material, raining down on young worlds in the form of comets and water-rich asteroids, may be a key process in forming a life-sustaining planetary surface environment.

SUPERNOVA 1987A

I n 1987, astronomers detected the explosive end of a massive star about 163,000 light-years away in one of the Milky Way's orbiting dwarf galaxies. Decades later, it has produced a mesmerising supernova remnant, shown here in remarkable detail by NIRCam. For the first time, Webb revealed new crescent-shaped features either side of the blue 'keyhole' structure that was previously known. A brightly glowing ring surrounds the core of the remnant, made up of gas that was ejected thousands of years before the explosion, which was subsequently re-energised by the intense radiation from the supernova.

M82

NIRCam

- F140M
- F164N
- F212N

Side-by-side views of the fascinating galaxy M82 offer us different ways to study its central region. Both images sharply resolve individual stars and clusters within the vicinity of the core, with unprecedented clarity. On the left, a combination of NIRCam's shortwave, narrowband filters have been used to isolate star-forming regions within the galaxy. On the right, longer wavelengths bring into view a complex process. Vibrant reds concord with dusty clouds that contain

polycyclic aromatic hydrocarbons. They are being energised and shaped by galactic winds, which result from the rapid star formation and high rate of supernovae erupting around the core. M82, which is about 12 million light-years away, is birthing stars at ten times the rate of the Milky Way.

SUNBURST ARC

A massive cluster of galaxies, roughly 4.6 billion light-years away, seems to be smiling at us in this NIRCam image processed by the author. In fact, the illusion is created by a gravitational lens, which is magnifying and distorting background galaxies. One in particular, which is about 11 billion light-years away, well behind the cluster, appears multiple times as its light follows the invisible curvature of space. Its brightest image forms the Sunburst Arc of the smiling mouth. Individual compact objects – stars or star clusters – are magnified into view, appearing as bright spots within the arc.

NGC 1672

NGC 1672 is a Barred Spiral Galaxy about 60 million light-years away. To find out more about this image, see page 48.

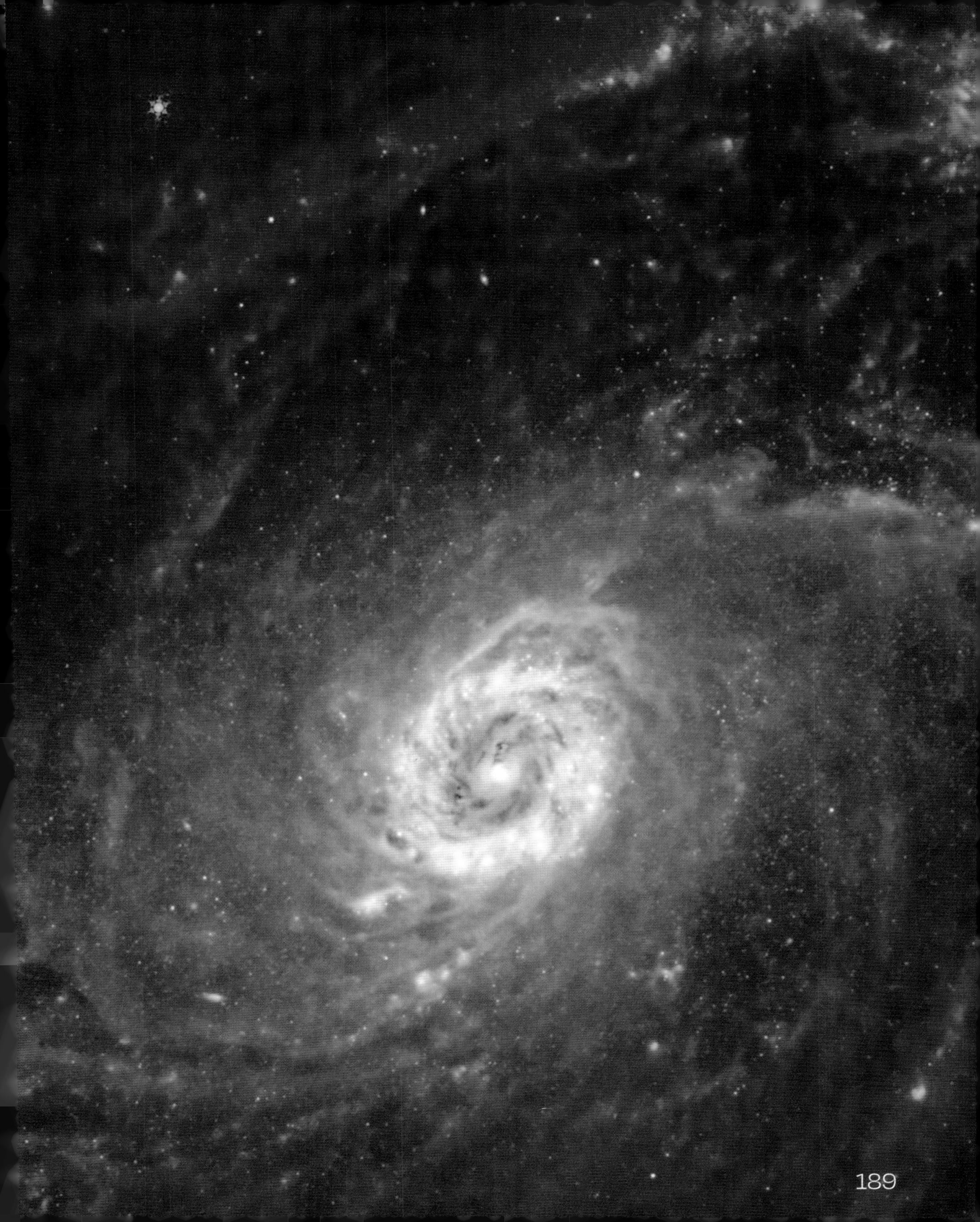

THE BEAUTY OF NGC 3603

The following four panes
are cropped from the same
NIRCam image and processed by
the author. They reveal fine details
in a portion of the star-forming
nebula NGC 3603. It surrounds
a massive super star cluster,
which lights it up with energy and
chisels intricate features out of
the gas and dust. In the left image,
a pronounced globule is seen
resisting the flow of surrounding
material. NGC 3603 is located
about 20,000 light-years away.

Four NIRCam filters were used to draw out subtle colours and textures sculpted by stellar winds. Red and orange hues are characteristic of abundant ionised hydrogen within the cloud. Pink is caused by a mixture of emission from the nebula and reflection of bluer wavelengths prominently emitted by nearby stars.

THE BEAUTY OF NGC 3603

Dark portions of the image appear to be free of gas. In fact, they are very dense clouds seen in silhouette, through which even NIRCam cannot see. Star-forming emission nebulae are vast, three-dimensional structures illuminated from the inside out.

NGC 3603 is the most massive nebula of its type yet known in the Milky Way. It is full of objects of considerable astronomical interest, which have resulted in extensive observation and study. Webb's instruments will shine new light on events occurring within NGC 3603 that have remained hidden from view until now.

THE HEART OF THE
ORION NEBULA
(SHORTWAVE)

Just to the left of the centre of this image is the glittering Trapezium Cluster, well known to amateur astronomers. It lies at the very heart of the famous Orion Nebula. This short-wave NIRCam image of the surrounding gas and dust offers a fresh look at a familiar region of the sky. Thousands of young stars are beginning their lives here, none more than a million years old. Some are surrounded by protoplanetary discs, sharply resolved by Webb's optics. In the upper centre of the image is a remarkable feature, which is detailed see page 198.

195

THE HEART OF THE ORION NEBULA
(LONGWAVE)

NIRCam

■	F277W
■	F300M
■	F335M
□	F360M
■	F444W
■	F470N

This alternative view of the core of the Orion Nebula was made using NIRCam's long-wave filters. These wavelengths delineate rich textures in the gas and dust. Purple represents ionised hydrogen, which floods the cavity hollowed out by the Trapezium Cluster. The stellar winds from these massive stars are producing erosion features such as pillars and cliffs in the green and brown filaments that surround it. Webb has detected numerous objects with masses below that of Jupiter, freely floating within the Orion Nebula.

EXPLOSION FINGERS

NIRCam

F140M
F162M
F182M
F187N
F212N

Around 2,000 years ago, a giant explosion occurred in the Orion Nebula. It may have been the result of a collision between two massive young stars. Today we see the aftermath in the form of these explosion fingers, which glow as energy from the detonation shocks the gas within the nebula. Molecular hydrogen produces the red colour, while at the 'finger tips' green and white indicate the presence of gaseous iron.

199

NGC 1566

N GC 1566 is a Spiral Galaxy about 60 million light-years away. To find out more about this image, see page 48.

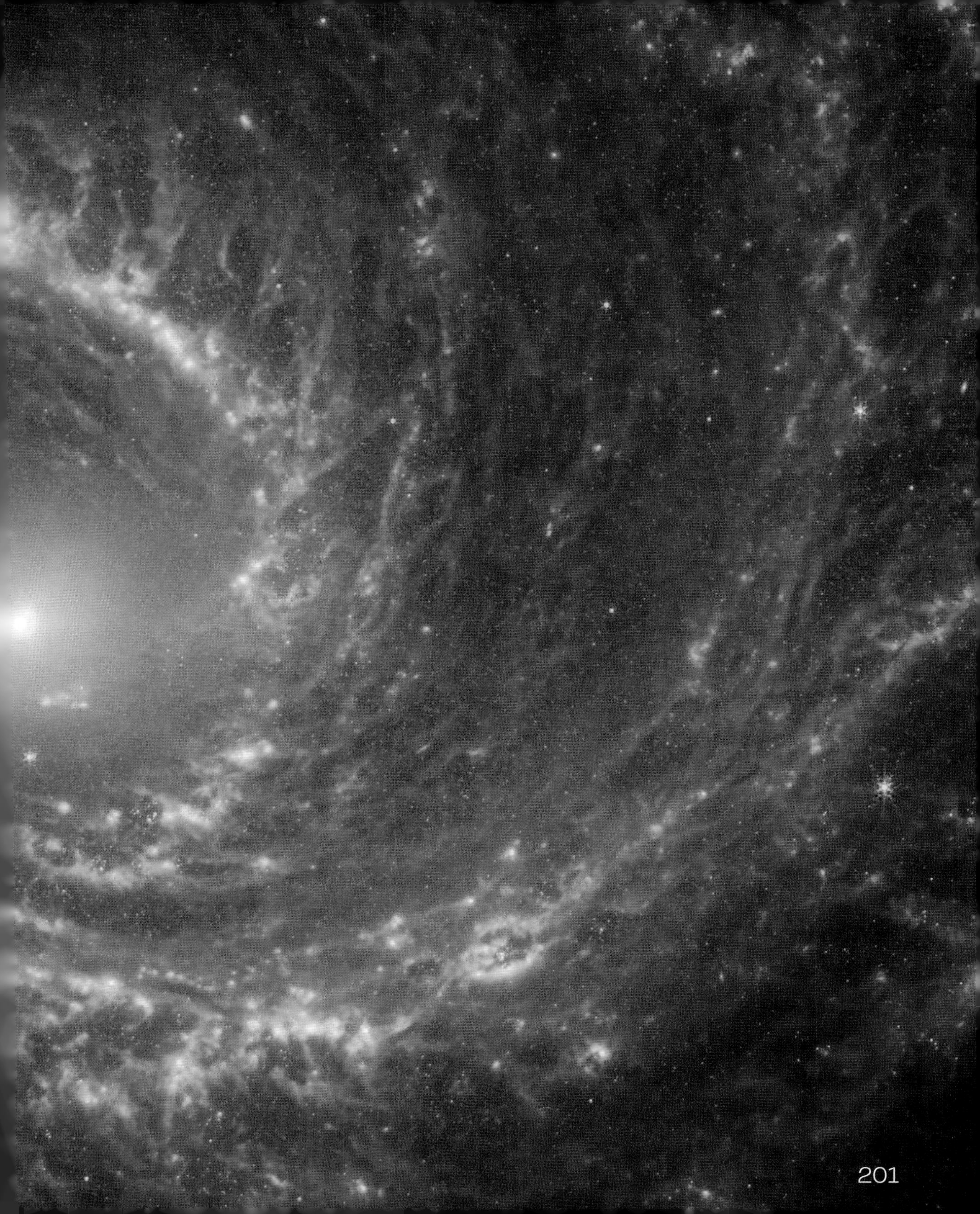

NGC 6822
(NIRCAM + MIRI COMPOSITE)

Both NIRCam and MIRI each contribute four channels to this strange and colourful image of NGC 6822, an irregular galaxy about 1.6 million light-years away. It is the nearest galaxy to the Milky Way, though at just 7,000 light-years in diameter, it doesn't rise to the grand scale of the famous Andromeda Galaxy, which is considered to be our closest significant neighbour. Nevertheless, NGC 6822 has an important history, being the first object considered to be extragalactic in 1925.

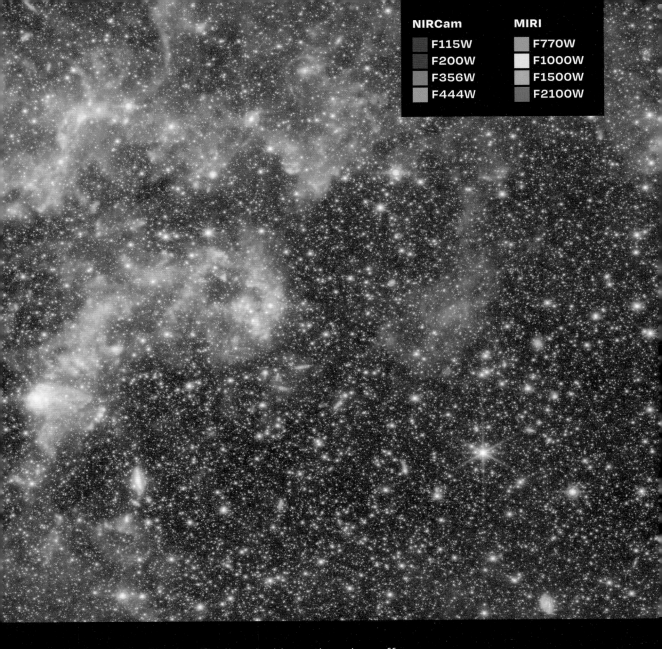

A century later, it's still scientifically valuable, as the galaxy offers astronomers the opportunity to study processes that occurred when the Universe was much younger. This is because NGC 6822 is currently sparse in metals (meaning elements heavier than hydrogen and helium) and this was the case everywhere before generations of stars fused and dispersed heavier atoms throughout the cosmos. NIRCam channels reveal dense populations of stars within the galaxy, whereas MIRI brings out swathes of dust and gas

NGC 6822
(SEEN BY NIRCAM AND MIRI)

Separate views of NGC 6822 from NIRCam and MIRI isolate populations of stars and gaseous regions, respectively. In the top image, NIRCam has delineated countless stars resolved by Webb's powerful optics. A bright blue globular cluster stands out in the lower left. In the bottom image, a variety of dusty regions glow at different temperatures, energised by radiation from stars. Warmer dust is represented in orange, and cooler dust is cyan blue. A few background galaxies appear green in this image, emitting infrared light from their own dusty clouds. Red and magenta regions represent active star formation, whereas the well-defined red ring is a supernova remnant left behind by a star that exploded long ago.

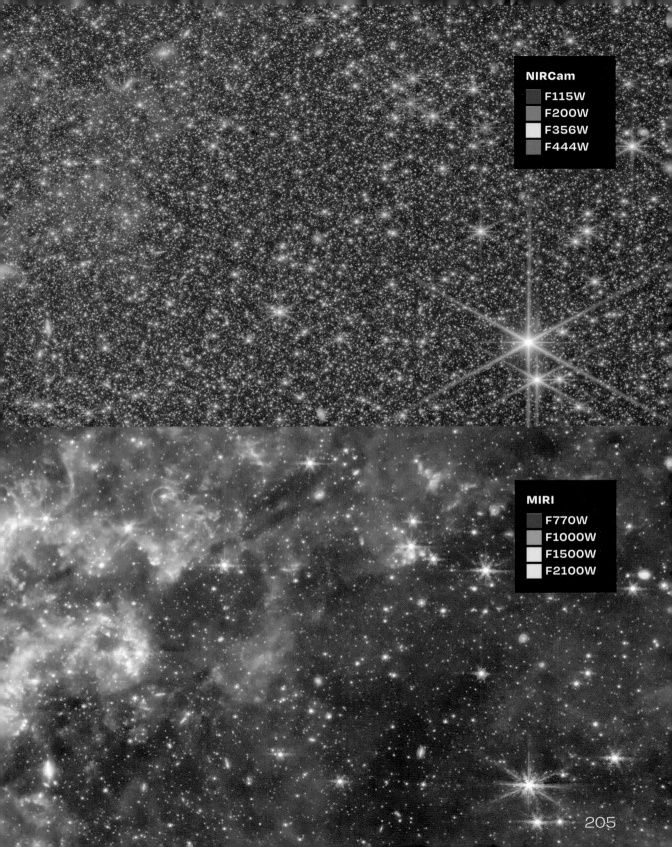

NIRCam
- F115W
- F200W
- F356W
- F444W

MIRI
- F770W
- F1000W
- F1500W
- F2100W

HH 797

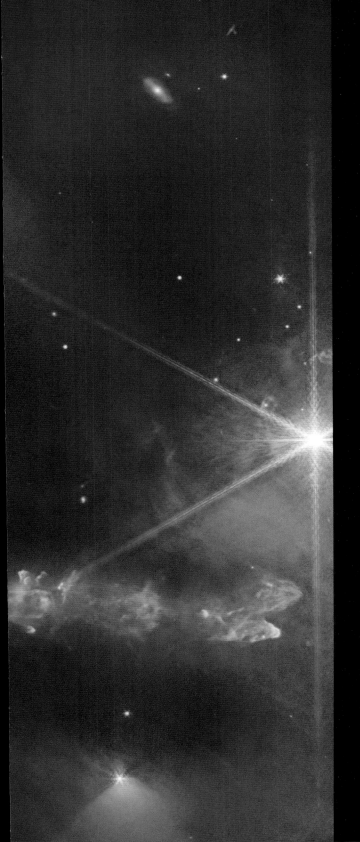

NIRCam

	F164N
	F150W
	F212N
	F210M
	F323N
	F335M
	F466N
	F470N

Herbig-Haro objects are challenging to observe in fine detail, but Webb's staggering resolving power is a perfect match for HH 797, shown here in its full glory. A shrouded, newborn star emits jets of gas that collide with a wider nebula, creating lobes that glow in the infrared. Molecular hydrogen and carbon monoxide are key components of the cloud that make the process observable, as they emit infrared light when they are excited by shocks that heat them to thousands of degrees Celsius. HH 797's outflows were previously thought to be rotating, exhibiting redshift and blueshift. NIRCam revealed for the first time that we are seeing two parallel outflows, indicating that there are two infants buried within the central cocoon – a binary protostar system.

THE COSMIC SEAHORSE

G ravitational lenses are known for the characteristic arcs they produce, as they warp and magnify the images of background galaxies. Here we something less common. A spiral galaxy's image has been smeared out into a large, triangular image. Astronomers call it the Cosmic Seahorse. The galaxy cluster in front of it has sufficiently brightened and magnified Seahorse enough to allow astronomers to study star formation within its spiral arms, despite its tremendous distance from us. Previously, Hubble analysed the same gravitational lens, helping astronomers better evaluate the observations from Webb.

209

THE AWESOME
POWER OF AN AGN

An Active Galactic Nucleus (AGN) is one of the most powerful types of astronomical sources. Even at a distance of 220 million light-years, the AGN at the head of NGC 7469 casts long diffraction spikes onto NIRCam's detectors. A supermassive black hole at the centre is feeding on vast quantities of dust and gas, emitting an enormous amount of radiation in the process. Being so bright compared to the rest of the core, this AGN has made its host galaxy relatively difficult to observe. Webb provides the resolution necessary to discern a bright ring of starburst surrounding the core, with a radius of just 1,500 light-years. Star-forming regions and turbulent pockets of hot gas close to the core have been studied for the first time, with the results showing that the AGN influences the region around it to a distance of several hundred light-years.

211

Five hundred million light-years away in the constellation of Delphinus, a pair of galaxies are seen locked in a merger that will one day see them form into a single object. Their individual morphologies have been highly disrupted by the gravitational forces that bind them, pulling apart spiral arms and throwing material into the intergalactic expanse. Among the chaos, long rivers of dust and gas, with active star formation, reach between the cores of the two galaxies, the activity motivated by the turbulent nature of the merger. MIRI makes clear the distribution of this gas and dust, which appears orange and red in the final image. NIRCam illuminates remarkably fine details within the bright blue spiral arms. Collectively, both instruments also bring to light many background galaxies in the Universe beyond.

M74 — THE PHANTOM GALAXY

M74 – nicknamed the Phantom Galaxy – looks utterly alien in this MIRI image, which uses an alternative palette to the one shown on page 98. Filaments of gas and dust, and the large cavities that separate them, are easily traced along the grand spiral pattern that sweeps out tens of thousands of light-years from the nucleus. Surrounding the core itself, unobscured by gas, is a large central cluster of older stars.

THE HORSEHEAD NEBULA
(NIRCAM)

This astonishing pair of portraits from both NIRCam and MIRI bring a region of the Horsehead Nebula, also known as Barnard 33, into sharp relief. This is a popular target in Orion for astrophotographers, but no ground-based images can compete with these. Far from Earth, Webb's pristine infrared vision draws out sumptuous textures in the top of the horse's head, with a sky full of galaxies seemingly overhead. Powerful nearby stars illuminate the cloud, which is, in fact, the end of a towering, light-years long column of neutral gas within the enormous Orion Molecular Cloud Complex.

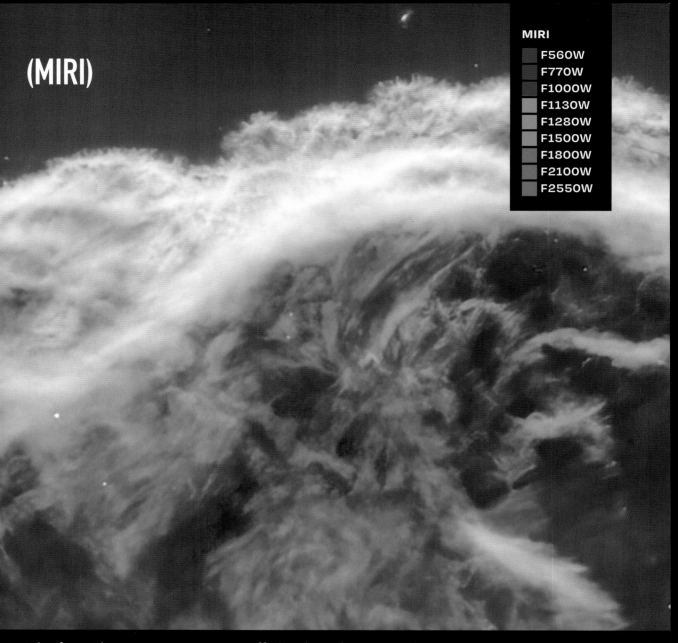

(MIRI)

It is of great interest to astronomers, offering them the opportunity to study the chemistry of interstellar material as it interacts with the radiation from stars. While the nebula is resisting erosion from stellar winds, having collapsed from a sparse cloud, it is likely that it will eventually disperse in about five million years.

In the background, Webb reveals a menagerie of distant galaxies never pictured before, some even shining through the nebula itself.

I ZW 18

NIRCam

F115W
F200W
F356W
F444W

About 59 million light-years away, a galaxy is exhibiting impressive displays of starburst, birthing swathes of new star systems in two different regions of its turbulent heart. Surrounding them, filaments and globules of gas – shown here in brown – glow with the energy of the intense radiation. I Zw 18 demonstrates that even an irregular dwarf galaxy can be a bustling star factory. Perhaps this burst of creation was motivated by interactions with another irregular galaxy, pictured at the bottom of the image.

The extraordinary power provided by NIRCam enables astronomers to determine the various types and ages of the individually resolved stars and clusters, in order to unravel this galaxy's history and address ongoing questions about star formation in the early Universe.

BIRTH OF A STAR
CLUSTER

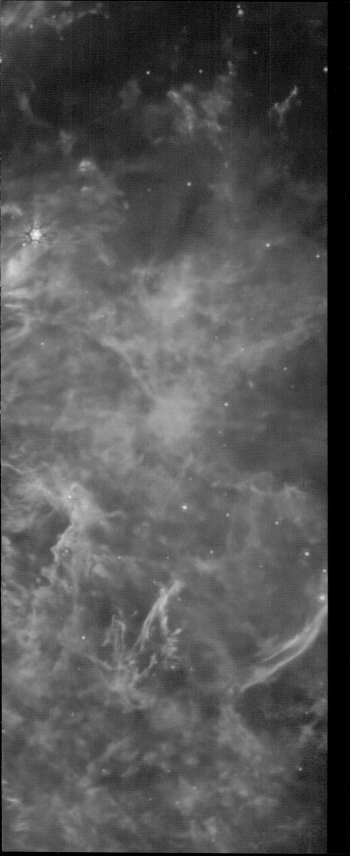

MIRI

■ F770W
■ F1000W
■ F1500W
■ F2100W

Buried within the Large Magellanic Cloud, a nebula complex called N79 is undergoing seismic levels of star formation. At the centre of this wonderfully turbulent fragment, brought into sharp relief by MIRI, the brilliant radiation of powerful young stars forms a series of diffraction spikes on the detectors. The surrounding cloud is chemically similar to star-forming regions that raged everywhere when the Universe was just a few billion years old, so it provides an analogue for piecing together cosmic history. Eventually, the star cluster at the heart of all this chaos will blow away the surround gas and dust, carving out its own space within the larger cloud complex.

IMAGE CREDITS

82–83	NASA, ESA, CSA, STScI, Janice Lee (STScI), Thomas Williams (Oxford), PHANGS Team, Joseph DePasquale (STScI)
84–85	NASA, ESA, CSA, Alyssa Pagan (STScI)
86–87	NASA, ESA, CSA, Alyssa Pagan (STScI), Nolan Habel (USRA), Laura Lenkić (USRA), Laurie E. U. Chu (NASA Ames)
88–89	NASA, ESA, CSA, Alyssa Pagan (STScI)
90–91	NASA, ESA, CSA, STScI, Janice Lee (STScI), Thomas Williams (Oxford), PHANGS Team, Joseph DePasquale (STScI)
92–93	NASA, ESA, CSA, Joseph DePasquale (STScI)
94–97	NASA, ESA, CSA, Alyssa Pagan (STScI)
98–99	NASA, ESA, CSA, STScI, Janice Lee (STScI), Thomas Williams (Oxford), PHANGS Team, Joseph DePasquale (STScI)
100–101	NASA, ESA, CSA, STScI, Webb ERO Production Team
102–103	NASA, ESA, CSA, Joseph DePasquale (STScI)
104–105	NASA, ESA, CSA, STScI, Joseph DePasquale (STScI)
106–107	NASA, ESA, CSA, STScI, Janice Lee (STScI), Thomas Williams (Oxford), PHANGS Team, Joseph DePasquale (STScI)
108–109	NASA, ESA, CSA, STScI, Alyssa Pagan (STScI)
110	NASA, ESA, CSA, András Gáspár (University of Arizona), Alyssa Pagan (STScI)
111	NASA, ESA, CSA, Mike Kelley (UMD), Henry Hsieh (PSI), Alyssa Pagan (STScI)
112–113	NASA, ESA, CSA, Brant Robertson (UC Santa Cruz), Ben Johnson (CfA), Sandro Tacchella (Cambridge), Marcia Rieke (University of Arizona), Daniel Eisenstein (CfA), Alyssa Pagan (STScI)
114–115	NASA, ESA, CSA, STScI, Janice Lee (STScI), Thomas Williams (Oxford), PHANGS Team, Joseph DePasquale (STScI)
116–117	NASA, ESA, CSA, Simon Lilly (ETH Zurich), Daichi Kashino (Nagoya University), Jorryt Matthee (ETH Zurich), Christina Eilers (MIT), Rob Simcoe (MIT), Rongmon Bordoloi (NCSU), Ruari Mackenzie (ETH Zurich), Alyssa Pagan (STScI), Ruari Mackenzie (ETH Zurich)
118–119	NASA, ESA, CSA, Mahdi Zamani (ESA/Webb), PDRs4ALL ERS Team
120–121	NASA, ESA, CSA, Feige Wang (University of Arizona), Joseph DePasquale (STScI)
122–123	NASA, ESA, CSA, STScI, Janice Lee (STScI), Thomas Williams (Oxford), PHANGS Team, Joseph DePasquale (STScI)
124	NASA, ESA, CSA, Matthew Tiscareno (SETI Institute), Matthew Hedman (University of Idaho), Maryame El Moutamid (Cornell University), Mark Showalter (SETI Institute), Leigh Fletcher (University of Leicester), Heidi Hammel (AURA), Joseph DePasquale (STScI)
125	NASA, ESA, CSA, Webb Titan GTO Team, Alyssa Pagan (STScI)
126–127	NASA, ESA, CSA, Ori Fox (STScI), Melissa Shahbandeh (STScI), Alyssa Pagan (STScI)
128–129	NASA, ESA, CSA, Steve Finkelstein (UT Austin), Micaela Bagley (UT Austin), Rebecca Larson (UT Austin), Alyssa Pagan (STScI)
130–131	NASA, ESA, CSA, STScI, Janice Lee (STScI), Thomas Williams (Oxford), PHANGS Team, Joseph DePasquale (STScI)
132–133	NASA, ESA, CSA, STScI, Klaus Pontoppidan (STScI), Alyssa Pagan (STScI)
134–135	NASA, ESA, CSA, Joseph DePasquale (STScI), Anton M. Koekemoer (STScI)
136–137	NASA, ESA, CSA, Alyssa Pagan (STScI), Jake Summers (ASU), Jordan C. J. D'Silva (UWA), Anton M. Koekemoer (STScI), Aaron Robotham (UWA), Rogier Windhorst (ASU)
138–139	NASA, ESA, CSA, STScI, Janice Lee (STScI), Thomas Williams (Oxford), PHANGS Team, Joseph DePasquale (STScI)
140–141	NASA, ESA, CSA, Zolt G. Levay
142–143	NASA, ESA, CSA, M. Barlow (UCL), N. Cox (ACRI–ST), R. Wesson (Cardiff University)
144–145	NASA, ESA, CSA, Alyssa Pagan (STScI)
146–147	NASA, ESA, CSA, STScI, Janice Lee (STScI), Thomas Williams (Oxford), PHANGS Team, Joseph DePasquale (STScI)
148–149	NASA, ESA, CSA. Processed by Dr. Mark McCaughrean